Lost Bastards

L Todd Wood

Lost Bastards
L. Todd Wood
ISBN: 978-1-943927-06-7
Library of Congress Control Number: 2016919178
Copyright © 2016 L. Todd Wood
All rights reserved worldwide,
Icebox Publishing, Westport, CT

Cover Design by EBook Launch.
Rear Cover Image by Dewey Mclean.

This story has been pieced together from the memory of Richard Carpenter's children in discussions with him regarding this period in his life, from Carpenter's memoirs, FOIA requests, and from Major Carpenter's friends. Some parts of the story were created to fill in the gaps in a way which would not distract from known facts. L Todd Wood was assisted by Richard and John Carpenter in the writing of the book.

To the children.

Eugene HannaLora (Laura Gene) Diehl October 6, 1946, Deceased

Sieglinde Margareta (Siggie) Carpenter December 29, 1950

John Robert Carpenter May 30, 1956

Ruth Anne Carpenter July 10, 1959

Diana Lee Carpenter b. 13 Mar 1961, Deceased

Richard William Carpenter December 1, 1963

In 1934 A.D., an emperor of the Ming dynasty of China, the Celestial Empire of the East, gave Korea the title of 'Chaohsien' meaning morning freshness. The title was most suited to South Korea because of its spellbinding natural beauty of picturesque high mountains and clear waters and its splen-did tranquillity, particularly in the morning which further confirmed the title on South Korea as the 'Land of Morning Calm'.

Times of India

"On the other side of every mountain was another moutain."

Lieutenant Colonel George Russell, Battalion Commander, 23rd
Regiment, 2nd Infantry, Korea

Preface

I didn't know what to think when I was contacted by the family of Richard Carpenter to write this book. However, the more I learned about the story, the more I was touched. I am humbled to have been asked to undertake this project. My only hope is that I have done the story justice. When you read this book, let it take you back to another time in America, almost seventy years ago. We had just fought and won the Second World War. We were tired of conflict and death. We just wanted to live in peace. There was much brightness ahead for America. We were the shining city on the hill.

But unfortunately, as we would soon learn, there was still evil in the world and it was on the march. It was called communism, which was just another word for totalitarianism. This evil would eventually kill several hundred million people before it was also vanquished. Actually, based on events we are seeing today, it wasn't completely eradicated. It only went dormant for a few decades, simply to reappear in our universities and media several decades later.

However, in the early 1950s, America just wanted to eat hot dogs, chase girls, and watch baseball. When evil did raise its head on the Korean Peninsula, America called for its warriors one more time. As always, they answered that call. They didn't cry. They didn't seek their safe spaces or demand trigger warnings. They just went, did their duty, died, or came home.

I'm going to open the door for you now to the Forgotten War. All I ask is that as you read, say a prayer for these brave men's souls. Honor them. On your journey to Hill 433, open your eyes, smell the gunpowder, hear the artillery, feel the machine gun fire raking the earth around you, feel the fear.

Honor them.

A rare "Double-Mustang"
of Distinction and Honor.

RICHARD L. "DICK" CARPENTER
Major
U. S. Army
Korea - Southeast Asia

Special Forces

A "Lost Bastard" who earned a Battlefield
Commission during the Battle of the Kumsong River
with the ROK Capital Division while surrounded
by enemy forces. After discharge, he re-enlisted and
became an Officer again, after OCS.

Prologue

It had been this way her whole life, as far back as she could remember. She was different from most others, and for some reason she could not understand, this was bad, in society's eyes. This difference made her disgusting, a criminal, a danger to German society. A danger that eventually Hitler decided needed to be exterminated.

Kunigunda (Gunda) Schuhlein (22 Jan 1929 – 16 Mar 1990) was the daughter of Johann Schuhlein (18 Apr 1893 – 27 Dec 1933) and Eugenie Amalie Jenieve Bauer (6 Nov 1894 – 25 Jun 1963). Gunda's father had been a German soldier in the first great world war. Her memories of him were vague. Really the only thing she remembered was that he was sick, always in bed, and always had trouble breathing or walking. Only later did she learn that it was due to the mustard gas in the trenches. He died young and left her mother and nine children destitute.

The fact that her father was a soldier kept the worst tendencies of German society away from her. If the authorities had known her mother was Roma, or a gypsy, she would have been killed with the others, shot, sent to the camps, whatever the Nazis felt like that day. Gunda's skin was dark, and became darker in the sunlight. The Roma were of Indian descent, having migrated up through central Asia and into Europe centuries before. She covered up as much as possible. She couldn't let her friends know the truth as the truth meant certain death.

She had heard of the camps. She had seen the horse drawn carts of the Roma herded by the police into a slum town near the garbage dump on the outskirts of town. She had heard of the forced sterilizations, the rapes, the killing, the stealing of anything these poor wretched souls owned, the death, the end of everything.

Her friends and family called her Die kleine Negeria, or in English, 'little niggar.' It was as if the Nazis stole her safety, her security, her self-confidence, even if they had not destroyed her body. Eventually the Nazis started classifying Roma based on physical characteristics as the church records did not list them as non-Aryan. All they recorded was that they were Christian, as all Roma were Catholic. Physical characteristics of pure, Aryan blood became vitally important. Gunda was constantly afraid her ethnic heritage would be found out. When the German youth were sent to work in the fields to support the war, she covered up as much as possible, deathly afraid of what the sun would do to her skin, as in making it darker.

She overcompensated to prove her Germanness. As the allied troops rolled into town at the end of World War II, she decided to try and help. Her friends had left her after hearing gunshots but Gunda spotted several abandoned German portable anti-tank rockets, or Panzerfaust, on the German side of the river after the bridges had been blown. She attempted to row the weapons across the river to German troops cutoff on the American-occupied side of the waterway. American forces opened fire, thinking she was a German soldier. She ducked and her German field cap, worn by all Hitler youth, fell off and they realized she was a female child. The boat was sunk. She stood on the shoreline, shaking her fist at the American invaders, her German blood boiling. She began throwing rocks at the Americans until she realized they were laughing at her, then walked off in frustration.

The other difference, that set her apart growing up was that she was poor, very poor. Hunger was a thing she kept by her bed every night, gnawing at her gut, interrupting her dreams of soup and bread. She never knew where her next meal was coming from. The house was always cold in winter.

She earned money as best she could. All of the children had to help. Starvation, death, and disease, were the alternatives.

She grew up during the Weimar Republic. As a child, the German Reichsmark collapsed as Germany attempted to pay war reparations with printed money. At the end of the First World War, the R-Mark was valued at four to the U.S. Dollar. Two years later, it was valued at over two million to one. The German treasury could not print the money fast enough to keep up with the rapid inflation.

The victorious allies saddled Germany with a debt, and a guilt, they could never repay. Germany sank into a horrific depression, as did the United States. History would forever cite the Weimar Republic as an example of a currency crisis that led to political upheaval. This is the primary reason Hitler came to power. Germans were poor, destitute, and defeated. Hitler promised to change all that.

Gunda became creative with ways to make money. Since her father had passed, her grandfather would take young Kunigunda around the beer halls to empty the ashtrays into a bucket. These were massive, ornately decorated drinking establishments which would fill with large German men drinking copious amounts of beer and smoking. Occasionally she would receive a pfennig (German penny) or two for her efforts and an occasional smack from a drunk. Inflation was raging in Germany. The printing presses could not keep up with the demand for cash as the value of a mark plummeted. A hundred pfennigs might buy the family a loaf of bread.

Her grandfather and Gunda would sort through the bucket to pull out the bits of tobacco and cigarette papers to recycle. She became very adept with her small hands picking out the shreds of tobacco. Her grandfather would then hand make cigarettes to resell.

The poverty and the teasing made Gunda strong. This, in addition to her German upbringing, created a force of will inside of the girl who was turning slowly into a young woman, and an attractive one at that. In later years, she would attempt to pass on this will to survive to her children. Gunda survived the war. She survived the Nazis as the child of a gypsy. She became a woman.

As the American tanks rolled into the city of Bamberg, German residents hung white sheets of surrender from the windows. Gunda hid in fear in the basement of the apartment where her family lived. She was sixteen. As the rumble of the tank tracks shook the house, she dared to pull back the drapes and take a look at the Americans, whom she had heard so much about. There were hundreds of them in all types of military vehicles. One soldier turned to look at her. She froze in horror! He was a negro! They all were negroes! Gunda's town had been liberated by American negro troops in an ironic twist of fate.

She guessed the negro soldier was over two meters tall! He had a wide nose and huge lips which protruded beneath the steel helmet. He was chewing a large wad of tobacco and turned to look at her as he walked next to the half-track and passed her house. The soldier saw her looking out of the window and smiled as he spit tobacco her way. He had huge yellow stained teeth.

Gunda shook in terror and ran to the interior of the house. Her siblings teased her terribly, "Gunda, the negro is coming back for you! Little niggar!"

Gunda didn't leave the apartment for almost a month as she hid from her deepest fears and the large negro soldier.

Chapter One

The years after the end of World War II were a perilous time in history for all of mankind. The threat of a nuclear confrontation was front and center as the Soviets developed the bomb and began constructing the horrific Iron Curtain in Eastern Europe. The Berlin Airlift of 1948 was a case in point as Stalin cut off the German capital from resupply and only a year long display of American airpower saved the day.

The world was deathly afraid of the spread of communism which had overrun two of the largest countries on earth, the Soviet Union and China. The fear in Southeast Asia of the red scourge drove the decision making in Washington.

The situation on the Korean Peninsula had developed in a strange fashion. After the removal of imperial Japanese rule, Russian forces occupied the territory north of the 38th Parallel. Here they installed a Stalinist leader who outdid the Soviet leader in his ruthlessness, as if that was possible. The People's Republic of Korea was formed.

American troops occupied the South per an agreement struck in 1948 with the Soviets. A reluctant America backed a dictatorship in the Republic of Korea. Neither government on the peninsula recognized the other and both considered themselves the legitimate authority of all Korea. This stalemate remained until June 25, 1950 when 75,000 well-trained and well-equipped North Korean troops invaded the South and wreaked havoc. American troops quickly joined the conflict, per a United Nations Security Council Resolution, but to no avail as in barely two months U.N. forces were forced down to the Pusan Perimeter in a desperate bid to survive, much less repel the North. The mood in Washington was dark indeed. Communism was on the march.

East Berlin
Russian Sektor
1952

Corporal Carpenter laughed so hard the beer came out of his nose. It had been a while since he had laughed, he realized, as he caught his breath and cleaned his uniform shirt where the German beer had spilled down his chest. His buddy next to him, Corporal Whitehorse helped in this process, dabbing his U.S. Army enlisted shirt with a rag from the bar. "You made a mess of yourself, Carpenter. This is going to be hard to explain to your little German wife!"

Carpenter took a moment to look around the bar. The Russian soldiers and the Americans were still dancing, the beer swashing around the wooden floor of the small beer hall in the Russian sector of Berlin. The bartender, unsure at first at the prospect of Russian and American troops drinking together, now was enjoying making extra money on a slow night. He kept the beer flowing.

The object of Carpenter's amusement, a large Russian female soldier, had buried his other friend's head in her bosom. Ransom's head literally seemed to disappear. "Ona sovsem sletela s katushek!!! yelled a Russian enlisted man (she might be going off the rails!). Carpenter lay back in his chair and slapped his leg. "I won't tell Gunda about this, Whitehorse. There are some things that are better off not said!"

Corporal Carpenter didn't want to think about his wife, his sick daughter, or his future. No, he just wanted to keep drinking. He needed a break; he had decided to let go of the tension. I had been building inside of him. It was killing him, the stress of it all. He downed the rest of the beer and signaled the bartender for another. *Tonight I will just enjoy myself and tomorrow I will pay the price and get back to the grindstone.*

Carpenter reached up and felt his face and realized it was still swollen. The alcohol had dulled the pain but it wouldn't make the acute puffiness go down. That would just take time, and maybe some ice. "Yeah, Corporal Riney really put a whoopin on you Carpenter. Good thing we bet against you. I knew you weren't ready for that fight. I could see it in your eyes. And, it was a grudge match at that! You were going to fight no matter what to settle the matter once and for all with those guys! Even if you weren't ready. But, you were doing it just for the money! I knew you were going to lose. So all ends well. At least we have plenty of cash for this little night on the town. I could tell you needed a night out too, my friend. That's what friends do, take care of their buddies. So keep drinking and by the way, that little honey pot over there has her eye on you. Don't worry, what happens here stays here," Whitehorse said with a smile.

Earlier in the day, Carpenter had fought a hard fight with one Corporal Riney, as they both were part of the Army boxing league and had traveled to Berlin for a fight against another unit. It didn't go well as the fight was called after the third round. Carpenter had the face to prove it.

But the fight wasn't his biggest concern. Corporal Richard 'Dick' Carpenter had married a German, a big no-no in the post WWII U.S. Army. To show its displeasure, the military went so far as to refuse his wife and daughter's medical treatment at the post medical facilities. Their dependent status was not recognized. To make matters worse, his youngest daughter, just a toddler, was very sick. This caused immense financial pressure on Carpenter to create cash flow to pay for very expensive antibiotics. Boxing was an easy way to make money and to get special passes from time to time. However, this time it hadn't worked out as planned. Luckily, his friend had been smarter and he had money just the same.

Instead of bedding down after the fight as instructed, they had illegally made their way through an apartment complex near Check Point Alpha to bypass the guards and security to the Russian sector, a move strictly against the rules, one that could get them court martialed. The Berlin Wall was not fully built yet and if you were determined, you could still cross rather easily between east and west. The American dollar went a lot further in the Russian-controlled area of the German capital, for beer or for a short time with a German working girl. In addition, Carpenter and his friends wanted to see Berlin, even the Russian sector. U.S. or Russian officers would be none too thrilled however to find American soldiers drinking in a Russian-sektor bar. The Americans took their chances anyway and stumbled upon a hole-in-the-wall beer cellar with the light on.

Earlier in the evening, several Russian, enlisted soldiers had surprised Carpenter and his friends in the hole-in-the-wall tavern with possible trouble to ensue. However, a few greenbacks on the bar did the trick to calm things down. Now they were all the best of friends, something military camaraderie and a few beers tends to manufacture. The Russians even pulled out an accordion, another soldier a harmonica. Soon everyone was dancing and having a good time.

Carpenter felt a hand on his shoulder. The girl from across the room who had been eyeing him had worked up the courage to take matters into her own hands. She was probably desperate for money as well. Carpenter let her stay. For the time being, he thought.

"Aaahh...horosha padla!! Americanitz!" cried one of the drunk Russian soldiers, which meant *very beautiful whore, American!* Carpenter nervously laughed again. These guys were fun, too bad we were basically enemies, but soldiers of all nationalities understood things like women and booze. For now, they were brothers in arms.

Suddenly there was a cry from the doorway where some of the Russian soldiers had been keeping watch for anyone in their chain of command who might ruin the impromptu party. "Uhodim, capitan idet!" *My captain! My captain! Let's go! Let's Go!"* the Russian soldier shouted!

All of the drunken Americans immediately shook off their German girlfriends for the evening, grabbed their coats and fur hats and headed out the back door into the cold night to sneak back to the American side. The Russian soldiers grabbed their weapons and went outside to meet their officer. The working girls were left frustratingly empty-handed.

Thirty minutes later the motley crew safely returned to Western Berlin without being detected. The men donned their coats and hats as the night chill began to sink in and the alcohol wore off. They stopped under a street lamp and Whitehorse lit up a cigarette.

"What the Hell do you have on Ransom?" Whitehorse yelled. Then a realization washed over his face and he howled, "You dumbass, you took a Russian's coat and hat!"

Ransom took off his hat and looked at the red star emblazoned on the forehead. "Damn, I thought this hat was too big. But the coat is sure warm!" he exclaimed.

Carpenter looked down and realized he had a Russian coat and hat on as well.

"What are we going to do now?" Whitehorse asked the others in the group.

"We sell them, for sure!" replied Carpenter. "We can make some nice money on the black market for those. We report ours stolen while on duty and get replacements."

"I have to admit you got it all figured out Carpenter," said Whitehorse.

"Now where to?" asked Ransom.

"To that little bar down the road, where else?" Carpenter responded. The men laughingly disappeared into the German night.

Hours later, and three bars later, Carpenter and the boys caught the red eye train back to Bamberg from Berlin. Boxing was now the last thing on his mind. He tried to catch some shut eye in the passenger car but it was no use. He would have to wait to get home before he would be able to sleep. The jostling and banging of the rail car simply wouldn't allow him to drift off. Carpenter's thoughts slowly came around to his family and the troubles he had tried so hard to forget the entire night.

Corporal Carpenter clumsily found his way home, the alcohol taking its toll on his senses. The sun was peeking its nose over the horizon. Luckily the streets were deserted at this time of the day as he neared his apartment. The recurring altercations with aggressive German youth had made his trips home from the post quite challenging as of late. He was relieved there would be no such combat this morning. His head would not have enjoyed it.

The pain of the recent beating with dehydration from a night of drinking combined to form a deadly pounding which he didn't want to share with the Hitler Youth at this time of day. Soon he reached the door to his building and made his way painfully up the stairs, wishing he had some aspirin and a glass of water.

Gunda met him at the door. She had been crying. Carpenter immediately felt guilty for spending the night out. *So selfish of me*, he thought to himself.

"Her fever is way too high! I've used all the medicine! We need more! The Doctor earlier said she was really sick. Dick, I'm worried! Do you have any money today?"

"Yes, I have a small amount for another couple days worth. It's okay. I'm home now." Gunda collapsed into his arms, exhausted after a night with no sleep taking care of Sieglinde. Carpenter carried her into the house and put his wife gently to bed. "I've got it now; go to sleep, Gunda," he said softly. She closed her eyes and moved no more as the exhaustion took hold of her.

Carpenter then went into the other bedroom and checked on his girls. Hanalaura, his adopted, older child, was sleeping soundly. Sieglinde, the two-year-old was wide awake and obviously in distress. He felt her forehead which was burning hot. She was wide awake but listless. Carpenter picked her up and cradled her in his strong arms, took her to the bath and bathed her in cool water to reduce her temperature. He would not sleep either that day and the stress returned with a vengeance, only now he had no sleep and a headache to go with it.

"Hang in there Siggie. Daddy's here now."

LOST BASTARDS

" *My earliest memory, circa summer 1955 Büdigen, Germany. Dad was always bringing soldiers from his unit home for a good German meal, music and card games. Mom loved to cook and was always the life of the party. After one field maneuver, he brought home a fawn whose Mother had been killed. Dad loved animals. Mom named him Willie after her brother Wilhelm who never returned from the war. Quarters had basement caged storage areas and ours was empty as all our furniture was military quartermaster. We nursed and cared for Willie until he grew horns and kept butting Mom, he was our first pet. The photo was taken at the military game preserve where we took Willie. We'd visit him after church on Sunday with his favorite treats we had collected during the week."*

Sieglinde Carpenter

Chapter Two

US Army Garrison Motor Pool
Erlangen, West Germany
October 1952

Carpenter squinted into the sunlight streaming across the tops of the trees as the clouds moved away and opened up a hole for the heat to burn through. It was seven o'clock in the morning and the air was crisp; he welcomed the heat from the sun's rays which bathed his face.

He stood at attention at the U.S. Army garrison motor pool, in Erlangen, West Germany, along with over two dozen other enlisted men. The men were organized into four sections. They formed up to wait for an announcement per the instruction of the command sergeant major; the content of this announcement they had no idea. But soldiers obey orders, so here they were, waiting. *It was always hurry up and wait in the U.S. Army,* Carpenter thought to himself. *I could use a cup of coffee right now.*

The sound of a U.S. Army jeep caught his ear, speeding in from the left, but he dared not turn and look to see who it was. A move like that at attention would invite a rebuke from the command sergeant and a week's worth of some unpleasant duty for sure. *We'll know soon enough.*

The jeep careened to a stop in front of the formation, an officer stood in the passenger's seat, his hands on the glass windshield. He turned to look at the enlisted men.

"At Ease! My name is First Lieutenant Sharp." he said to the men and then waited for them to relax. "I am told you are the best of the 1st Infantry Division. I am here to select as certain number of soldiers for a special mission. To be honest, I don't need all of you. So, over the next training cycle, you will prove to me which section is going to

27

be the best at using those crew weapons..." He pointed towards a line of armored vehicles, smiling. The vehicles were M16 halftracks holding quad 50 caliber anti-aircraft weapons and fully tracked M19s holding dual 40mm cannons.

"*And*, to make things a little more interesting, the two sections that clobber the others in qualifications will get Christmas leave! The losers get to detail out all the vehicles and weapons for Christmas inspection."

There were shouts of excitement amongst the soldiers, who soon would be in competition for a coveted pass off the post for the holidays. Carpenter sensed a chance to make money and whispered to the man in formation in front of him, the same corporal who had beat him senseless in the recent boxing match. "Riney, you want to lose 20 bucks for top section?"

Riney, his hands folded behind him, flashed Carpenter four fingers, raising the bet to $40.

"I don't know," responded Carpenter in a hushed voice. He paused to think it over. "Okay. Forty bucks. It's a bet." Riney flashed a thumbs up, then the middle finger.

Staff Sergeant Mike Kowalski and Sergeant Al McCoy gave orders to their squads who mounted the vehicles and began preparing them to leave the motor vehicle compound. Carpenter and Whitehorse, under Kowalski's command, fell in line behind the Lieutenant's jeep as the convoy moved out. The jeep led the convoy to the firing range.

"You are either an idiot or desperate to make that bet with Riney," Whitehorse said to Carpenter as they maneuvered the vehicle down the post gravel road.

"I need the money."

"Now Ransom and I need to work our butts off for you. You will owe us Christmas dinner for this!"

"Deal done. Just help me get the cash. Siggie's really sick and desperately needs it! I'm scared."

Lieutenant Sharp stood in front of the now even further reduced number of enlisted men, once more formed up, this time on the firing range at the post. Two days had passed of intense weapons training and firing competition. Carpenter was especially nervous. Christmas with Gunda and the girls was imperative. Siggie was recovering but still needed a great deal of tender loving care. *I really need to win this thing.*

The officer's sergeant handed him a clipboard. He scanned the data and then addressed the men. "The results are in. And the results are clear. Kowalski's squad is number one and McCoy is number Two. Those guys get Christmas leave. The others get to GI the vehicles." Carpenter's group cheered while groans emanated from the losing sections. Carpenter relaxed. As they were dismissed to return to quarters, Carpenter jabbed Riney in the back and said, "Pony up Riney, you owe me forty bucks!"

"Yea, yea, okay, here you go." Riney dug into his pocket and produced the cash. "You won fair and square. But I'll be looking to win that back soon, Carpenter!"

"Any time, you know I'm up for it."

As they walked to the vehicles the command sergeant major called out, "You ladies enjoy your holidays. When you return in two days, we get to work. And, you find out what all of this is about. So enjoy your time off and get some rest. You're gonna need it!"

Whitehorse turned to Carpenter, "I don't like the sound of that."

"No, neither do I," Carpenter admitted out loud.

Ransom, unaware of their conversation, ran up and grabbed both of them around the neck. "Let the fun times begin! We're headed to your place for Christmas right Carpenter?"

Carpenter had met his wife Gunda two years earlier while on a weekend pass with a buddy to visit the Fränkische Schweiz (Bavaria's Switzerland) which was a beautiful mountain area near Bamberg.

They got bored quickly with the stunning scenery, and decided to take the train back to Bamberg that Sunday morning. On the way to the train station, they went into a café for breakfast. The place was nearly deserted, with only a few older persons present. Right after they sat down, near the door, another older woman and what Carpenter viewed as a young girl came in. He made somewhat of a snide comment about not finding any women except seventy, and seventeen, not thinking that anyone spoke English. Someone approached their table, and he looked up to see the young girl. She spoke in English and said, "I beg your pardon, I am not seventeen, and you are very rude." With that she and the older woman left without ordering.

Carpenter fell in love at first sight. The two soldiers finished their breakfast, and went to the Bahnhof to catch the train back to Bamberg. The two women were on the platform, and boarded the same train as the two American soldiers. Carpenter did not see her again until he got off the train in Bamberg. He approached her there, and tried to make amends. *Man, this girl really has got a hold of me for some reason.* But she would give him nothing but a cold shoulder, and drove off in a Taxi.

After that, and for months, he looked all over the city for her, and was beginning to get depressed, thinking that he would never see her again. Having formed a friendship with his boxing coach, who was a Sergeant and lived off post with his wife in Quarters appropriated from the Germans, Carpenter was invited to his home for dinner. It was some kind of a special occasion, and they were having Goose for the evening meal. He nearly fainted when he walked through their door, and discovered that Gunda was their maid. He was overjoyed, and spent a considerable amount of time that evening getting under her feet, as she tried to go about her work. She finally put him out of his misery, and agreed to a date the next day. As an aside, the goose as very rich, and made him sicker than a dog; he threw up several times on the way back to the Barracks. The very next day he started to court Gunda in earnest, and it was only a couple of months until she said yes. The two love birds tried to get permission to marry, but neither the Americans, nor the Germans would allow them to do so, without waiting months, and doing a significant amount of paperwork.

In early February 1950, Carpenter took a furlough, and the two drove over to Bern Switzerland and got married on February 6th. Then they took a honeymoon trip through Northern Italy, Southern France, and into Spain. They wanted to see Paris, so they headed north through France. That was a big mistake, as the French were very rude, because Gunda was a German, and traveling on a German Passport. They spent only one day and night in Paris, and drove back to Germany. They never returned to France.

I was a very young child and my towering father was always a beacon of safety and security. It was Easter and I was on an Easter Egg hunt struggling through grass that was tall to me with dozens of others about my age. From his glorious height he would softly call my name then point and I would run/stumble with my large basket until I found a colorful egg and claim the prize that was now mine.

A time later I was risen from my small height and posted on his great chest of honor and held firmly in his firm arms. He told me he was proud of all the eggs I had found & I was happy in the moment. Then he pointed at a young girl with her mother. We knelt. I remember him saying something but I was shocked/confused when he instructed me to take two of my own eggs and place them in her basket. I felt it was wrong, they were mine! But I loved my father.

Then I remember looking at the girl's face and seeing sudden happiness. I looked up and saw my father was smiling, and looking around the girl was smiling, her mother was smiling and I found myself smiling too.

Again the great pleasure of being lifted unto Olympus, being in warm and secure arms and knowing all was well.

John R. Carpenter

Chapter Three

After their marriage, Dick and Gunda had returned to army life. He was frequently restricted to the post or placed on temporary duty in the field. It was around this time that the war on the Korean Peninsula broke out. Carpenter was assigned as a clerk to the battalion sergeant major and was also frequently the battalion commander's driver. During these long periods of restriction to base or deployed on exercises, Carpenter took the time to become extremely knowledgeable on weapons, particularly crew-served anti-aircraft systems. He spent hours and hours alone on at night or on the weekends, learning about his new passion. This personal quest and hunger for weapons expertise is what led to him being 'volunteered' for the secret mission to Korea. It was also during this period, while Private Carpenter was frequently deployed, that Gunda was raped and became pregnant. A poor, single, German girl in American-occupied Germany was an easy target. It was the second time she had been taken advantage of. She broke the news to him, probably one night over tears, when he returned from one of his frequent temporary duty assignments. After processing the information, Private Carpenter made a decision. He loved Gunda deeply and would accept the child as his own. The two of them made a pact. They would never tell the child the truth. She would always be 'their daughter.' American troops were always coming and going in 1950s Germany. Siggie would never have the chance to even find out who her real father was. Whether or not this was the best course of action, would remain to be seen. After DNA tests in later years, Siggie found out she was not Carpenter's biological child. It caused her great distress and a feeling of 'not belonging.' Gunda and Dick went to their deaths without revealing the truth.

In November of 1950, due to his weapons expertise, Carpenter was sent on temporary duty to a new weapons school in another part of Germany.

"Gunda had a daughter from a previous relationship with an American soldier. When the war ended, American forces occupied Bamberg. Gunda was just 18, and trying to survive. She stayed at home, and looked for work, which was nearly nonexistent for a young woman. In early 1946 she met an American soldier, who got her drunk, and then took advantage of her. They subsequently established a relationship which lasted for about six months, until the soldier shipped out. She named the child after her Grandmother. When I first met her she was four years old, a lovely, but very spoiled child. At first I did not know how to relate to Laura, I loved her mother, but Laura did not want anything to do with me. So we established a very strained relationship with very little interaction between us. However, I began to feel affection for her, and was able to slowly improve the relationship. When her baby sister (Linda) arrived, that broke the ice, and our relationship steadily improved, and my love for her grew. She became my own daughter."

Richard Carpenter, Memoirs

Gunda gave birth to Siggie two months premature on December 29th, 1950. She was born with the assistance of a Hebamme (midwife) in Gunda's parents apartment as the American authorities still did not recognize Gunda or her daughter Hanalaura as Carpenter's dependents. Carpenter and Gunda had been living with her parents in Bamberg, Germany. Carpenter, who had not been home in weeks, was granted leave by his section commander who took pity on him. He arrived home on New Year's Day in 1951. The doctor was called after her birth because Siggie was born so premature. No one thought she would survive.

However, the doctor and the midwife rigged a homemade

incubator and Siggie miraculously survived. She steadily grew stronger.

"Linda was originally named Sieglinde Margareta in German. The German authorities would not accept Linda Margaret as her name. It was not on their list of acceptable names. Later when she, Laura, and John became naturalized citizens we changed her name to Linda Margaret. Linda was named after a German Aunt Sieglinde, and an American Aunt Margaret. We thought it important to recognize the two different cultures that had come together in her birth."

Richard Carpenter, Memoirs

Carpenter remained on temporary duty at the weapons school and rarely saw his family. In October of 1951, he was returned to his original post in Bamberg and reunited with the girls. He was overjoyed.

"During my absence the clerical position I held in the battalion had been filled by another soldier. They needed a clerk in the regimental personnel office, and I got the assignment. My job was to do all the paperwork clearances for soldiers whose tour was ending, and who were returning to the United States. It was strictly an 8:00 am to 5:00 PM, and I loved it. It gave me I lots of time to spend with my family.

"It didn't take long for me to realize that many soldiers were not able to get a full clearance to leave, because they had cars registered in their names, and had not disposed of them. Most of the cars were being held in Germany Repair Shops for payment of the repair bills. So I worked out a system with a friend of mine, Whitehorse, (a mechanic) to buy the cars for a pittance. The soldiers were happy. They got to go home. The Repair Shops were happy. They were able to dispose of an American Automobile that was

costing them storage, and to which they could not acquire title. Whitehorse and I were happy as we were making quite a bit of money from the system. We also made a friend, who was a military police desk sergeant. He gave us a temporary registration for each car which was good for 90 days, but did not show up on the official records. That way, when we sold the cars to other soldiers, it appeared that title had been transferred directly from the soldier who had gone home.

"Another part of the deal was the black market. Each American Registered Vehicle was able to receive a 100 gallon gas coupon book each month. There were several American Gas Stations at the Exchange Facilities where we could get gas in exchange for the coupons. The books cost $10.00, or 10 cents a gallon. Whitehorse owned a converted jeep, that had a 100 gallon gas tank. As soon as we received the temporary registration, we would buy a gas coupon book, and then fill his gas tank. With a full gas tank we would then visit the German repair facility where the car was located, and strike a deal. Gas on the German economy was very expensive, over four dollars a gallon. In every instance we were able to swap gasoline to pay the repair bill, and take possession of the car. Whitehorse would then get the car into selling condition and we would park it near the regimental exchange complex with a "For Sale" sign on it. We had no problem in disposing of two or three cars each month and this resulted in a real financial bonanza for each of us, even after we paid off everyone. I would estimate that during that time I was getting $500.00 each month for my part in the endeavor. I wish I could have kept it up forever. I stayed in this assignment for well over a year, until the deal fell apart with the departure of my buddy, and the military police desk sergeant., who rotated back to the States. I was still a private first class. I had not really put any serious thought into getting promoted, and making a career of the service. But, I had become addicted to the extra money and was looking for a way to improve my salary. To be promoted to corporal, I had to attend the regimental noncommissioned officers academy. So I volunteered, and in May of 1952 I was assigned to *the academy. It was a four week course and I had no problem passing it. On graduation, I was promoted to Corporal."*

Richard Carpenter, Memoirs

Upon leaving the academy, Corporal Carpenter was assigned as a section leader, weapons platoon, 60mm mortars, A Company, of the 1st Battalion. He again spent weeks in the field, away from his family. To rectify the situation, he tried out for the regimental rifle team. With all of his weapons expertise, he was accepted easily and was assigned to this new special duty assignment with all kinds of free time to spend with Gunda and the girls. Carpenter was nothing, if not creative and persistent, to make sure he was the best father possible, in his own way.

The earliest memory of my father—we had just come back from Germany. We were watching the station wagon being lifted off the boat. It was a sight to see. Then the cable broke, needless to say down came the car!

We ended up spending the week in New York. We saw the sights including the Statue of Liberty. The parks I remember were all street like, Made out of asphalt. After a week, we were on our way to California. So many stops along the way. Historical; Little Big Horn, Gettysburg, Paul Bunyan. It was fun as a kid. Siggie usually sat in the back because she got car sick. She thought it was great; she didn't have to pass snacks or drinks that mom packed. John, Rich and I along with mom and dad. Dad mainly drove at night and slept during the day. He always got a motel with a swimming pool. Kept us kids busy! We saw relatives along the way. My aunt Annie, my mother's sister. Grandma Braaten, dad's mom, she lived on a farm. I remember her flowers being so tall or maybe it was me being so short. I am not sure which it was. LOL. We made it to San Diego in 1966. We stayed with my Uncle Toby, Aunt Edith, Marjorie, and Chuck until my parents bought the house on Twin Lake in La Mesa. Precious memories of family get togethers, birthdays, thanksgiving and Christmas. The many times trying to find Rich who always seemed to find the pond and him being naked as a jaybird.

Ruth Carpenter

After the first two months of the conflict on the Korean Peninsula, U.N. forces, 88% comprised of American troops, were pushed down to the Pusan Perimeter; the conflict looked lost. However, General MacArthur landed new forces in a daring amphibious assault at Inchon and encircled the enemy, forcing them back up to the Yalu River. At that point, the Chinese entered the war, supported by the Soviet Union. U.N. forces retreated south to the 38th Parallel and the fight became a war of attrition.

December 1952
Parade Ground
Erlangen, West Germany

Carpenter once again stood at attention. There was no banter in the air this early morning. The holidays were over and they were once more formed up waiting on the Lieutenant. *No one is laughing this morning. No, this time it is serious.* He shivered in the cold German morning, in spite of his field jacket.

There were just shy of twenty of them now from the original

group of soldiers. They were the ones that had proved their skills and impressed the officers. However, a couple other men had been added to their motley crew, a medic and radar specialist, upping the obvious danger level for the mission. None of the men were oblivious to these new developments. All that was left was for the Lieutenant to drop the other shoe, and tell them where they were headed, although most of them already knew.

Once again the jeep roared up to the formation and screeched to a stop; Lieutenant Sharp was again standing in the passenger seat, for effect. He had a smile on his face.

"I got a bad feeling when an officer smiles like that," Whitehorse whispered to Carpenter.

"Yea, what the Hell have we got ourselves into?"

The Lieutenant spoke, "Gentlemen. From this moment on you are restricted to post. You are not to leave quarters without permission. You are not to communicate with anyone outside this post. Is that understood?"

"Yes, Sir," was the uniform response from the enlisted soldiers of the detachment.

"You men have been selected, due to your superior skills with the crew-served, anti-aircraft weapons we have been training on for the last few days. You are the best of the best. Your country needs you and I will be proud to lead you into battle. Yes, we are going into combat. We are going to Korea. We will be inserted into the middle of combat operations under cover as an anti-aircraft unit. We will be testing a new technology, battlefield radar, a technology that will give us an edge in the war, and against our enemies all over the world. We obviously don't want the Chicoms to know we have this capability to pick them out and target them on the battlefield, so this is an extremely sensitive mission. We are to test and perfect this technology in combat conditions. This technology could literally turn the tide of the war. I trust you will do your duty and I know I can count on you..."

Sergeant First Class Quincy, the detachment sergeant, looked over at Whitehorse. Although they were all friends, Quincy had put his sergeant hat on and now was signaling the other enlisted guys in the formation to shut the fuck up. Quincy was a career man who saw combat at the Battle of the Bulge. The squad sergeants Kowalsky, who was also a combat veteran, and McCoy took the hint and made a facial expression to the other men to keep it down while the Lieutenant was speaking.

"I don't like this one bit," Carpenter said under his breath. He had the look of death on his face, like all the blood had drained from his head to his feet. "What the Hell is he stressing combat for? I've got a sick girl at home I have to take care of! What does he mean we can't leave the post?"

At that point, the Lieutenant started introducing new members of the team. First there were the radar operators who were trained on the new equipment. Then attention turned to a man who had joined the rear of the formation. He had a red cross armband, a helmet with the red cross, and was carrying a medical kit. "This is Private First Class Casper. He will be our combat medic."

That was when the realization hit the squad that they were going into harm's way. Carpenter wasn't worried about himself. He was worried about his family. *How will I get money to the girls? How will I even get word to them that I am shipping out?*

The Lieutenant kept talking. As he talked, Carpenter noticed armed guards surrounding their tent compound. "None of you can leave this area," he heard the Lieutenant say. *This is fucking serious,* he thought to himself.

"As we assembled with other volunteers, we were immediately isolated in a tent complex, and guards were placed around our compound. The next morning a signal corps major, and an artillery lieutenant briefed us on the assignment, and told us it was Top Secret. We were not to discuss it, under any circumstances with anyone outside of the assembled group. The assignment was to form a test unit to explore the potential of a new

anti-vehicle, anti-personnel field radar system under combat conditions. As a cover, in addition to the radar, we were being assigned the equipment, and weapons of an anti-aircraft battery. This included four half tracked vehicles. Two mounted with quad fifties, and two mounted with twin forties. We were to become proficient with both the radar, and the weapons."

Richard Carpenter, Memoirs

The lieutenant continued, "Radar was the ultimate wonder weapon at the start of the last World War. The Brits used it to channel their limited air power to where the Germans were flying into England. By the end of the war, it seems that everyone had radar. Even the Krauts had a radar which could locate Russian tanks coming at them ... little good it did them. Since then radar has become more sophisticated and smaller. As you know we have radar that can see and track incoming artillery rounds for counter-battery fire. In the States, I have helped develop a new generation of radar that can see soldiers in the field at night, in fog, and even if they are camouflaged. We can see them coming and hit them before they know we are there. We will be conducting our training at the old German tank range at Tagelow near Erlangen ... When we are ready, I will be proud to lead you men to Korea for a 3 month tour for combat testing. Thank you. I thank you. Your country thanks you. Now you will be sent to your individual squads for more in-depth training on the equipment. Sergeant?"

"Dismissed!" said Sergeant Quincy.

Carpenter walked briskly up to Sergeant Quincy from behind as he walked back to the trucks, then tapped him firmly on the shoulder. The man turned around to face him. "Sergeant! ... I have less than a year left and I am married. And…"

Quincy cut him off, "I read your file Carpenter. According to the Army, you are not married to any *Nazi* yet."

A look of rage washed over Carpenter's face. He tensed up as if he might punch Quincy, which would be an offense worthy of a court martial. Whitehorse stepped forward and put his hand on Carpenter's shoulder. Ransom and Kowalski also moved closer to restrain Carpenter if need be.

"If your dependent application gets approved by division, then we will deal with it. And *only* then. You should have gotten more than a slight slap on the wrist for fraternizing with a ... German national. You are officially single just like every swinging dick out here. You volunteered, now don't come whining to me soldier," Quincy adds. He then storms off, angry himself.

Kowalski steps in front of Carpenter, blocking the view of the noncommissioned officer walking away. "Don't worry, all AA units in a combat zone are stationed behind the battle lines. That is Army SOP. And don't forget you get combat pay!" He smiled, trying to calm Carpenter down. "And besides, the war may be over before we even get there."

Ransom spoke up, "Who volunteered for Korea?"

"None of us did," replied one of the radar specialists.

"I sure the hell didn't."

"Me neither. I didn't volunteer for anything. I was assigned. This is bull shit," added other members of the detachment.

Kowalski, one of the squad NCOs jumped in, "Enough ladies. You either volunteered or got volunteered. Shit happens! Deal with it. Now, get over it."

McCoy added, "Time to get on the trucks. Grab your gear and let's go. Move it!"

Our training started immediately with artillery officers, and NCO's from the division artillery units stationed in Erlangen training us on the vehicles, and weapons. The Major, and the Lieutenant did all of the training on the radar systems. We were all experienced soldiers, and it did not take long for us to become an cohesive unit.

We trained day, and night until mid-February 1953. Then we had a test of our skills; we were sent to the field in a nearby training area. As part of aggressor forces, facing an attacking infantry regiment, our mission was to detect all the enemy movements, and classify them for the aggressor forces commander. To accomplish this we had to split up into four teams. Each team centered upon one radar set, and one half track vehicle, protected by an infantry platoon.

We passed the test with flying colors. In every instance of troop, or vehicle movements, our teams were able to detect them, in particular at night. Our reports to the aggressor commander allowed him to shift forces, and meet every movement of the Regiment, For the regiment, it was a disaster, for us a celebration, In early March we were advised that we had completed our training, and were being shipped to another location. We had to box up all I of our equipment, and personal items, and get them ready for the move. It only took us a couple of days, and we were alerted to move in mid-March.

Richard Carpenter, Memoirs

Tangelow Test Range
West Germany
1953

Carpenter walked out of the tent carrying a rolled-up sleeping bag and headed for the deuce-and-a-half pulling a trailer as the detachment prepared to move out. Whitehorse was busy packing the unit's gear. The guards around their compound were especially tense this evening, as if they knew the day was not normal and change brought risk. They kept an eye out for any odd occurrences around the compound, as ifs the North Koreans themselves were on the prowl in Germany. "You know, we've been guarded here like virgins in a whorehouse," Carpenter lamented. This has been the tightest security I have ever seen. All my efforts to send a letter or call have been thwarted. I really miss my wife and kids. My wife is probably thinking I just up and disappeared by now. Sergeant Quincy found out that my dependent request still is in limbo. So, if anything happens to me in Korea, my wife and kids will get nothing! The worst thing is there is not a f'in thing I can do about it. Gunda needs money for Siggie and I'm f'ing stuck here. Unreal. She could die if she doesn't get medicine!"

Sergeant Kowalski yelled out, "Carpenter, Whitehorse! You guys ride with the equipment! Hurry up! All aboard! Next stop is Frankfurt's Rhein-Main Air Force base!" The MPs helped Carpenter and Whitehorse into the back of the covered truck and took seats beside them. Slowly the convoy moved out onto the autobahn to transport the men and equipment to the airfield for the flight eastward. Whitehorse took out a pack of cards to try and take Carpenter's mind off the situation. It was no use, he looked as sad as a puppy separated from his mother for the first time. The military policemen were nervous, looking out the back of the truck as they caressed their M-1 carbines with rounds in the chamber. *I'm tired of this top secret crap*, Carpenter thought to himself as he tried to visualize his family and what was going on at home. *I have no idea if I will ever see them again.*

Suddenly the truck downshifted and began to slow down. The massive vehicle then drove onto the side of the road and eventually stopped. The two younger MPs in under the tarp in the rear grabbed their weapons. The MP sergeant smiles at all of the men and says, "Relax, stopping as scheduled."

"Carpenter! Get your butt out here! Now!" yells Sergeant Kowalski.

Carpenter scrambled out the back of the truck and saw Kowalski and the MP sergeant grinning like they just took candy from a baby.

Kowalski handed Carpenter a wad of money. "If anyone asks, we had bad fuel," Kowalski says. He then grabbed Carpenter and turned him so he could see towards the front of the truck. There, standing next to a vehicle parked on the side of the road was his wife, Gunda.

"It's medicine money for the baby. You got five minutes, Daddy," said the MP sergeant with a grin on his face.

Carpenter rushed forward and grabbed Gunda in a warm embrace. He didn't want to let go. "You're going to war?" she asked incredulously.

Carpenter handed her the wad of money and quickly took off his watch and ring. "Sell these if you have to, but take care of Siggie. I'll be back for these, if I can, I promise!"

Gunda buried her head in his shoulder and weeps. Carpenter pulled out his wallet and emptied it of money as well. Then he hugged her one last time.

"Almost brings a tear to your eye, doesn't it" said the military policeman from afar near the military vehicle.

"Yes, that it sure does," replied Kowalski.

Carpenter forced them apart and then sadly walked back to the truck and hopped in the back. Gunda was left watching them as they drive away, stunned.

However, Carpenter was overjoyed and humbled at the camaraderie of his fellow soldiers. He felt like a giant weight had been lifted off his shoulders. In fact, he was speechless, and didn't utter a word until they were loaded on the plane at Rhein-Main.

We were trucked to the Frankfurt Rein Main Air base, where we spent a couple of days in isolation. Not able to go anywhere unescorted by senior NCO's, or our Lieutenant. In late March we boarded an air force cargo airplane, a big Globe Master. Our first stop was Gander Newfoundland, and the next Fort Sill Oklahoma. At Fort Sill we spent a couple of days learning about some improvements that were being made to our radar equipment. Then we again boarded the same air force cargo plane. We stopped in Alaska long enough to refuel, and then flew directly into Kimpo Airfield in Korea. We all pretty much knew where we were going, but we were not told until we left Fort Sill. We arrived in Korea on the 1st of April (April Fools Day) 1953.

Richard Carpenter, Memoirs

Chapter Four

Carpenter and the unit arrived at Kimpo Airfield in Korea and immediately were driven with their equipment by truck to the north side of the airfield. There were tents waiting for them and they bedded down for the night after establishing guard duty for the equipment and a security perimeter for the entire group. The first thing they noticed was the smell, and the hulks of destroyed aircraft which had been pushed into the grass and left to rot in the hot sun.

Kimpo Airfield was positioned west of the city of Seoul and lay in a vast plain at the foot of the mountains to the north. Rice paddies surrounded the flat surface of the field and were fertilized with human waste. Kimpo had changed hands several times during the war and General MacArthur inspected the troops at the field in 1953 after the U.S. Army had pushed the Chinese back north to the 38th Parallel. The men of Carpenter's detachment still were of the opinion that their little 'excursion' into the warzone would be brief and unexciting. They were very wrong.

The men were inserted into the Korean conflict at the behest of the U.S. Army Security Agency. This unit was responsible for signal intelligence collection as well as electronic countermeasures. It was the successor to Army intelligence operations during WWI. The ASA reported to the National Security Agency or NSA. In 1977 the unit was merged with the U.S. Army's Military Component to form the United States Army Intelligence and Security Command.

Corporal Carpenter and the detachment were once more in formation outside their tents at Kimpo Airfield. It was morning but the sun was already beaming in an attempt to burn off the overnight cooler temperatures. Carpenter had his field jacket on and was as comfortable as he could be considering he was about to go to war. The Lieutenant once again walked out to brief them on the coming day's events.

"Gentlemen, we are being redeployed up on the line. We won't be fully engaged in combat as our position will be approximately one mile south of the forward line of troops. We will be bedded down with a Republic of Korea, or ROK, anti-aircraft artillery unit of the ROK Capital division, assigned to the American corps. That sector has been fairly quiet and it should give us a good chance to test out the equipment. However, we will be in a war zone so take care of your buddy next to you and be prepared for combat at all times with a moment's notice. That is all, load and prepare your gear for the drive north!"

The convoy to transport the detachment's men and equipment consisted of two trucks with trailers and a jeep with a trailer as well. The progress up into the mountains was slow as the roads were damaged from the war and not well maintained. "I don't want to break an axle in one of these ditches," Ransom said to Carpenter as he steered one of the heavy trucks.

As the convoy drove through a village, the children came out to beg. They saw only the very young and very old. Many rice fields were being planted. Only occasionally did they see the debris of war. They stopped at an armed checkpoint of very serious looking ROK troops and the convoy's papers were checked.

As they progressed closer to the MLR (Main Line of Resistance), they passed through a deserted bullet ridden village and paused at a control point before starting the drive up a hill at the orders of Korean MPs. A truck with wounded and dead ROKs was seen coming from the front. The wounded ROKs glared at the Americans as they passed the stopped American convoy. The men's attention was then directed off to the side of the road by loud voices where several Koreans were standing blindfolded. With three quick orders in Korean, a firing squad readied, aimed and fired. As the convoy started up the hill, a ROK MP Officer walked the line of those shot and calmly shot each one again in the head.

Eventually they arrived at the ROK position after dark and

quickly unloaded their equipment and supplies into a large bunker on the top of Hill 433, which is now located in the Demilitarized Zone or DMZ, which separates North and South Korea. The ROK troops showed the men a portion of the bunker complex and they bedded down for the night. Machine gun fire and the rolling thumps of artillery could be heard in the distance. The Lieutenant cautioned the men again to keep their weapons at the ready. Guards were posted and the men spent their second night in-country and their first of many nights on Hill 433.

"The next morning we got a good look at the unit, and the position we were in. It appeared to be just over a mile from the front lines. With binoculars, and the artillery spotter scopes we were able to trace much of the friendly lines, into no man's land, and in one valley a little bit of the Chinese lines. The position looked pretty good; the batteries' four half-tracks were well dug in, and well spaced. A series of trenches connected everything together. The position area looked to be about half of a football field in size. In the center of the position, and on the South side was the large bunker where we had spent the night. The only other Americans there were a forward air controller team, an Army sergeant, and two other enlisted men. The FAC team's Lieutenant had been wounded and evacuated, and was never replaced while we were there."

Richard Carpenter, Memoirs

The men did not sleep well in their new surroundings. Carpenter woke early and made his way up the trench from the bunker to the top level, then stepped into the fresh, morning air. The fifty square foot space was a communications, command and observation area. Only about three feet of the structure protruded from ground level, and was very heavily sandbagged. The Lieutenant was observing the enemy positions over a mile away, using binoculars. The sound of fighting could be heard, although it was less intense than the previous evening. It seemed both sides needed sleep before the killing began again.

"I can see a lot of activity on the Chinese side." He handed the binoculars to Carpenter and pointed north east. "See, to the right of that valley, along the ridge, you can see enemy activity. The rest of their lines are hidden."

Carpenter raised the glasses to his eye as the sun broke over the ridge and attempted to locate what the LT was talking about. Finally he spotted what Lieutenant Sharp had seen, his first look at the enemy. There were a few dozen men scurrying around a fortified position. It looked as if they were attempting to move a gun emplacement. "Yes, I see them," Carpenter replied. He handed the binoculars back to the junior officer.

Carpenter took the moment to look around. It had been very dark when they arrived the previous evening and now in the daylight, he wanted to get a look at his surroundings and understand the lay of the land. As he turned and walked back down the trench, Whitehorse and Ransom met him halfway.

"Quite the operation they have going here. Did you take a look downstairs?" asked Whitehorse.

"No, I didn't. Let's do that now. What do you say?" Carpenter responded.

The men continued down the tunnel which led from the upper observation deck and soon were in the second level of the bunker. There were several rooms hewn out of the rock which were used primarily to house troops. The facility was obviously an abandoned mine which had been converted into a bunker complex. The actual underground system was huge and included four levels. The American detachment was boarded in one of these rooms which included enough cots for all of the GIs. Many of the men were still sleeping and Carpenter, Ransom, and Whitehorse quietly walked past them to the lower level. Electric lights glowed weakly above them and provided enough light to make the trek downward into the depths. Soon they exited the tunnel into another cavern which was not as refined as the upper levels but functional. It mainly consisted of a food storage area and a makeshift kitchen. Bags of rice lined the walls and the smell of dried fish permeated the space.

Carpenter almost stumbled over a very narrow railroad rail and he blurted out, "An old mine tunnel!" A few ROK soldiers ahead of them looked back curiously and realized the American was talking to himself and not to them. The 3rd level was under the boarding area and on the low end he could see where the tunnel had collapsed. "That must have been the original entrance to the mine." Carpenter walked slightly uphill toward the back end of the chamber. Old mining equipment and an ore car were long abandoned in the opening. He saw more sacks of rice and dried fish. As he turned to explore further, he heard dripping water and was amazed to find a cistern of water built near the collapsed end of the tunnel.

Whitehorse responded, "Someone planned this very well." The Americans looked up and saw water dripping from above into the cistern.

"Must be a couple hundred gallons," said Carpenter.

"Three hundred at least," said Whitehorse.

Ransom dipped his finger into the water, "It is cold and it's clear."

Carpenter and Whitehorse were examining the cistern when Ransom noticed a small Buddhist shrine and statue against the wall. He picked up the statue. "Look what I found! A fat little man!" Beside the statue was a small bowl with pieces of rock candy inside. Before Carpenter or Whitehorse could say anything, Ransom picked up a piece of candy and popped it in his mouth. A sudden cry of anger in Korean from across the room alarmed the men. A group of armed ROK soldiers suddenly confronted the three GIs, hands on their pistols slung around their waists. The officer in charge said in Korean, "Stupid American. They have no respect. He is defiling the shrine."

"Idiot!," said Carpenter, realizing what happened. "Take the candy out of your mouth and put it back in the bowl! Now!"

Ransom, now fearful, did as he was told. However, the Korean officer showed no sign of backing down. His hand gripped around the pistol and Carpenter stepped forward to diffuse the situation. "Take a few pieces of gum and place it in the bowl!" he further directed Ransom.

"My gum? Seriously?"

"Do you want to get shot? You know what a shrine is right? You just stole their offerings and defiled it! You idiot!"

Ransom suddenly grasped the enormity of the situation and let out a meek, "Oh." He slowly took the gum packet out of his pocket and placed a couple sticks in the bowl. The Korean officer slowly took his hand off his pistol.

"Now step backwards," Carpenter commanded. "And give the little fat man a bow!" Ransom did as he was told.

"Okay," grunted the stocky Korean. He turned and walked away back towards his troops.

"Everyone back to work," Carpenter yells out. "It's over!"

The Korean soldiers across the room glared at the Americans but slowly went back to their task. "Nice job, Ransom!" Carpenter murmured under his breath as the Americans ventured further down the tunnel into the depths of the bunker complex.

They soon arrived at the fourth and lowest level of the facility. It was another crudely hewn room blasted out of the rock and contained stacks of ammunition, literally filling the entire cave-like structure. The chamber was vast and extended hundreds of feet in both directions. The doorways were supported with wooden beams.

"I guess we've seen it all now," said Whitehorse.

"Yes, I guess we have," responded Carpenter. "Let's go check out the equipment while we have some free time." The men made their way back to the top of the facility that the Koreans had fondly named, *Steel Ridge.*

Chapter Five

The geographical area we were in was a long fairly flat ridge line running mainly east and west. The ridge tapered to the north and west toward what appeared to be a small river in the valley. To our northeast was a small cliff that dropped off into another valley. To the south east was a truck trail coming up the mountain, that ended at our position. It was apparent that the Koreans were going to build a small air strip on top of the ridge line. It had been graded for about a thousand yards, and in most places was 50-100 feet wide. The East end of the strip was right in front of our bunker, and the west end also ended in a bunker complex.

There was a large stockpile of pierced steel planking, literally thousands of them. After we had been in the position for about a week, with no one working on the air strip, some of the planking appeared in overhead cover built above our trench lines. Evidently the Korean commander of the guard unit thought it was a good idea, and if anything happened he could blame it on the Americans. So it did not take long before all of the trench lines, and the spaces around the halftracks were covered. In fact some of the cover constructed over the area was three, four, and even five planks deep. In short order the position was organized, and fortified. We had pretty good observation during the day, so to maintain the secrecy of our radar units we only mounted them as it got dark each day, near the four corners of the complex. We could track all of the vehicle, and foot movement in the area with specific identification as to what it was. The radar units worked perfectly, Several times we were able to identify Chinese reconnaissance patrols, and direct our units to intercept them.

Richard Carpenter, Memoirs

A few days had passed and Carpenter was now focused on the crew-served weapons that had been placed at their location as cover for the radar unit testing. After all, weapons were his specialty. The four anti-aircraft halftracks were dug in and secured before they arrived. Currently his attention was focused on one of the quad-50 half tracks, which was already fortified in the compound rather professionally. "The Koreans did a good job," he commented to Ransom who was helping clean the weapon."

"Yes, they did. A few rookie mistakes but we will teach them."

The Americans spent quality time with their new brothers in arms, teaching them how to operate the weapons systems and more importantly, how to maintain them. The experience bonded the two groups of allied men together like only a soldier's experience could. Soon they were slapping each other on the back, talking smack about women, and any memories of defiling the Buddhist shrine were long forgotten.

The Koreans did good work at everything they understood, unfortunately many of them did not understand the weapons they were assigned. They had received basic training as infantry, and simply put in the unit to build up its manpower. So, we set up a training team. We worked with the experienced Koreans (mostly NCO's) and with interpreters to instruct, and built up the confidence of four gun crews on their specifically assigned weapons, in a hands-on training program. It worked very well. In a short period of time most of the Koreans became quite skilled at handling, cleaning, and firing the weapons. The Korean battery had about fifty men, and there were an additional one hundred plus men in the logistical supply point, and guard units. With our complement of 28, and the FAC team of 3, there was right at two hundred men assigned to the position.

Richard Carpenter, Memoirs

Whitehorse and Ransom stood silently underground as they watched a Korean soldier painting a large mural on one of the bunker walls near their sleeping quarters. The man was extremely talented and the shape of a large, white tiger pouncing on his prey was slowly coming into focus. With their gear stowed and their quarters secured, there was time to kill during the day when they were not on watch. The two men were happy to have something to take their minds off the creeping boredom. The reality of combat had not sunk in yet as the war was still a mile or more from their position. As they watched, Carpenter and another ROK soldier, Sergeant Cho, walked down the ramp into the bunker. The Korean painter spoke hardly any English at all.

"Hey Dick, this guy is pretty good, don't you think?" Ransom asked Carpenter when he came into sight.

"Yea, he sure is," Carpenter replied.

"But why do you think the tiger is white?" Ransom continued. "I thought they were orange."

Sergeant Cho, who spoke better, but broken English, spoke up, "The white tiger is bigger! More fierce! Better fighter! Our unit is called the White Tigers!"

"Do you mean the 2nd Battalion or the regiment?" Carpenter asked.

Cho turns to the painter and asks him in Korean to draw the name of the regiment in English. "The man now draw name for our regiment!" The painter proceeded to write the number 26.

"The 26th Regiment? We came from the 26th Regiment!" Ransom exclaims.

Cho's face lights up. "Yes, yes, 26th Regiment is White Tigers!" He points to the Korean letters the man was now drawing near the tiger on the wall. "This Korean word for white tiger!"

Carpenter jumped into the conversation. "Our 26th Regiment is called the Blue Spaders."

Cho looked confused. "What is that?" he asked.

Whitehorse added, "In World War I, in France, our regiment fought with the French "Blue Devil" Division. They taught us how to dig trenches with shovels. The French call shovels spades. To honor them we call our Regiment Blue Spaders."

Cho slammed his fist into his other palm in front of his face which made a loud smacking sound. "It is good! White Tigers come together with blue eyed shovel to dig good fighting position and win battle! Special blue eyes to see enemy!"

"I'm going to go tell the other GIs we're all from the 26th!" Ransom said and briskly walked up the ramp to where the other Americans are gathered.

"Ransom is a dumbshit sometimes, but he is a good soldier," Whitehorse said to Sergeant Cho.

"That is the word. Dumbshit. By shrine we called him a dumbshit for bad thing he did to, ah, religion thing, ah …how you say in English?"

"The shrine?" Carpenter asked. "The little fat man?"

"Yes, to shrine. Shrine to Buddha. But dumbshit learns!" Cho cracked a wide grin. Carpenter and Whitehorse looked at each other and laughed loudly.

"Yes!" Whitehorse exclaimed. "Dumbshit can learn!"

Time seemed to drag by, we kept busy, but had a lot of free time. The big topic of conversation was when the Armistice would be signed. It appeared to us that the war was practically over, except for the shouting, and we wanted to go home. Unfortunately about the middle of June the Chinese launched a large offensive against the Korean Units in our area. The fighting went on for weeks, but the lines mostly held, and the Chinese made little headway. There was one area on the ridge line directly north o

f our position that changed hands several times, but was in the hands of the
Chinese at the end of the month.

Richard Carpenter, Memoirs

The next couple of weeks on Hill 433 saw the Americans training the ROK troops on the radar system and testing the unit under battlefield conditions. The position of Hill 433 became more dangerous as Chinese patrols made multiple attempts to penetrate the perimeter and continuously probed the hill's defenses. Carpenter and his team repeatedly calibrated the radar equipment and began using the system to vector ROK troops towards the Chinese as they advanced. The technology worked perfectly, allowing the Americans to pick out the enemy at night and lay down fire on their position. U.S. Air Force ground attack aircraft were frequently in the area and the Americans became astute at using this firepower effectively by spotting targets on the ground and relaying data to the American planes. With his love and knowledge of weapons, Carpenter was overjoyed to develop new tactics and techniques which could possibly turn the tide of the war. However, the real effectiveness of the system was yet to be tested. Carpenter and his men just didn't know it yet.

Carpenter sat next to Sam Smith and Rodriguez, looking at the radar screen. The glow of the presentation cast a green glow on the faces of the men in the darkness. "See the ROKs are waiting for the bogies to come to them. Almost there! Action!" says Smith. A few shots are heard in the distance.

"They got them," added Rodriquez. "Now the more important question is were they more lost ROKs or another Chinese patrol?"

"I don't know. Only a few shots," said Carpenter. "Maybe just lost ROKs. I will find out. The patrol will come up near the road and enter near halftrack 4. I'll be back, see you in a bit." Carpenter made his way down the trench to the other side of the bunker. He got to track 4 and found Riney and his radar techs on duty.

"Hey, look what the cat drug in boys!" Riney exclaimed.

Carpenter ignored the comment. "Is that patrol heading back in?"

"Yep, they got somebody," responded Unger. "They should be here in a few minutes."

"Last night they brought in a wounded ROK. The poor guy was lost in the dark. They treated him special. Lucky for him we spotted his movement on radar and vectored the patrol to him," said Da-Vid.

"Well, I'm going to go see who they got," said Carpenter.

"I'll go with you," added Riney. Riney and Carpenter headed to the opening in the wire near the road coming up the hill. There they waited with the ROK soldiers on duty for the patrol to come in. Soon they heard the outpost challenge the patrol but couldn't quite hear the password. However, they heard troops coming up the road. A radio telephone jingled and the ROK officer answered and said something in Korean. Riney & Carpenter watched the scene unfold.

The ROK officer did not look happy. The patrol quietly came up and squatted just outside the defensive position. A ROK NCO drug a ROK soldier with his hands tied behind him up to the officer.

"They got a deserter. They didn't bring in the other one, which means he is dead," whispered Riney.

The Korean officer screamed at the man the patrol apprehended. "Stand up you son of a dog! Name and unit!"

The man mumbled, "Liu Park, I am from the 11th Company. I demand…"

The officer struck the man in the face. "You demand? You are a deserter! Your dog of a friend was a deserter! He shot himself. Now I will shoot you!" The officer reached for his sidearm.

"Damn, he's going to shoot him!" exclaimed Riney.

Carpenter stepped forward. "Now hold on fellas! We can't just shoot him! He needs a trial. Besides, maybe he knows something!"

"I just tried him and found him guilty. He knows nothing!" responded the ROK officer. He looked at the deserter. "Do you know anything?"

"I'm just tired of fighting. I want to go home," said the man softly.

The officer became enraged. "We all want to go home! But you must fight to have a home to go home to! You are filth! You disgust me!" The officer raised his pistol again to shoot the deserter. Carpenter stepped in between them.

"Call Lt Col Kim!" he said. "Let him decide!"

The officer looked at Carpenter and then at the deserter in disgust. Then he directed a nearby soldier to call the commander, Lt Col Kim. "Please tell our commander to come! Tell him the Americans interfere with justice!" The ROK officer returned his pistol to his holster. The deserter crumpled on the floor crying. He was not more than nineteen, Carpenter guessed. Riney and Carpenter looked at each other in shock. A short time later, Lt Col Kim arrived with the messenger.

"What is the delay here?" he asked angrily.

"The American here says this dog filth may know something. He doesn't. The American said give filth a trial. I found him guilty of deserting. When I proceed with justice, the American says to get you. I was told to be nice to the

Americans, so I sent a messenger to you."

"Thank you," the colonel responds in Korean. Without a further word Kim drew his pistol and fired one shot into the back of the head of the deserter, killing him instantly. Kim turned toward Carpenter with a cold look in his eye.

"Justice must be swift for us. Discipline and duty must prevail. If you ever become a leader, you must learn this." Kim holstered then stalked off, leaving the Americans speechless. Carpenter and Riney looked at each other incredulously and then slowly made their way back to their positions.

Our position also suffered as the Chinese artillery worked through our area. Because of our bunker and reinforced trench line, casualties were few. We did suffer the loss of one halftrack, and its 40 mm Korean gun crew, when it received a direct hit. The twin forties, and the half track were a total loss, but we rebuilt the position under, and around the vehicle as a strong point in our defenses. The twin forties presented a higher profile than did the quad fifties, and it was only a short time later that we lost the second vehicle and guns. We thought it was only a matter of time, and our other vehicles with the quad fifties would be knocked out. So we dismounted the fifties, and put them all on ground tripods. Then we reinforced with steel planking, and sandbagged all of the vehicles as strong points. The Fourth of July came and went without us celebrating at all. Chinese patrols, and infiltrators were all over the place, and we had our hands full trying to track them. Plus more and more of the Koreans deserted the front lines, and tried to sneak to the rear. Most of them were caught and summarily executed.

Richard Carpenter, Memoirs

Chapter Six

The next morning, Detachment Sgt Quincy lined up all the enlisted men in ranks on the end of the air strip. The cool, Korean air was sharp with the morning dew as Construction continued on the small field at the other end of the makeshift runway. "Lt Sharp wants you to have at least 30 minutes of calisthenics. Sgts McCoy and Kowalski will lead." Quincy then started to walk off back toward the bunker.

A voice from the ranks called out to Quincy, "Sgt Quincy! Aren't you going to lead us?" yelled Riney.

"I need to do some supply ordering," he responded, while attempting to sound authoritarian. The men softly snickered at Quincy's response.

"Another full ammo reload and another full load of 10-1 rations. Oh, that is hard to order!" added Brenner. The men nearby laughed.

McCoy then took over, "Okay. Enough! Let's get this over with!" The men then proceeded to complete their daily dozen callies with lackluster performance, so much so that they finished early and were dismissed. Carpenter, Whitehorse and Ransom slowly walked off back to the bunker.

"Our ROKs are happy with us sharing our American food and cigarettes," said Ransom. All the men smiled.

"Good as money," said Whitehorse. The men stopped when they heard a series of whistle tweets.

"That sounds like a code," said Carpenter.

Carpenter looked around and saw ROK Sgt Park and the other ROK NCOS in a small group. One repeated the whistle tweets. Off duty ROKS soldiers from the bunker complex came running and started to form in ranks. A large group of ROKs were seen marching from the western end of the hill in a four column formation.

"This is different. A parade?" asked Ransom

The ROK NCOs gave commands and the ranks opened to double arm distance. The ROK soldiers were intent and not haphazard like the Americans were. The ROK leader began with a bow to Sgt Park. Then stretches commenced.

"What? Those are stretches," declared Ransom again as he tried one. "Those are good stretches."

Carpenter, Whitehorse and a few other Americans who were leaving returned to watch. Carpenter tried the stretches also. Several ROK NCO were correcting some of the ROK soldiers in their techniques.

"Oh. Ah. That feels good," Carpenter commented.

Suddenly, with a shout, the ROK soldiers went into a martial arts ready stance. Basic 1-2 punches were done in mass.

"What is this? Some type of callies?" said Sam Smith.

"It is like boxing! But different. See the stance? It is solid. And the punches are jabs," Carpenter said knowingly.

Ransom moved closer to the forward ROK position then tried to imitate them. Several ROKs soldiers looked disgustedly at the American. Carpenter saw this and moved forward correcting Ransom's stance and showed him how to do it. Then Carpenter started focusing on what the ROKs were doing and followed their example. Trying some of the punches and after looking at Ransom, Carpenter looked back towards the Koreans and was surprised to suddenly see Sgt Park in front of him. "Punch upward. Not straight. Aim here! Or here! indicating solar plexus or throat Drive punch to spine!" Park commanded.

Carpenter and Ransom corrected and after a few reps, Park nodded his head and moved on. Carpenter was breathing hard and sweating. The workout was getting to him. He notices most of the ROKs were in better shape than he was. Finally it was over and the men came to attention. Automatically Carpenter and Ransom did also. Sgt Park said something in Korean and the men shouted back. Then, they gave him a slight bow. As Sgt Park started heading back to the bunker area, Carpenter and Ransom gave him a slight bow. Park stopped and returned it. The slight bow was as an equal. As Park passed by, the

Americans were watching in a newfound respect.

"That was a workout! said Ransom as he collapsed to the ground.

"That was fun!" said Carpenter. The Americans continued to participate in the ROK, daily, martial arts drills and slowly their technique and skills improved with time.

Carpenter stared at the chessboard, lost in deep concentration. He tuned out the world and focused on winning, staying a few steps ahead of his opponent. Private First Class Scott Casper, the detachment medic, tried to do the same but it was no use. Carpenter was simply a better chess player and they both knew it. It was only a matter of time before the game was over. A few more moves and Carpenter said quietly, "Check mate." They both shook hands over the board in the bunker common area. Suddenly they both were startled by a Korean voice.

"Mind if I play the winner?" said Lt Col Kim, the battalion commander. Both of the Americans stood in respect for the rank and the man. Kim sat and looked up at Carpenter. "Sit down and let's play" he said in perfect English.

"Yes, Sir!" said Carpenter and he sat, pulling up the chair to the table.

The board was set up and Carpenter shook two pawns in his hands, then palmed each of them and presented two fists to Kim in order to choose his color. Kim chose white and the game began.

Several rounds quickly commenced as the pawns were taken down. However, soon the game progressed to the more important pieces on the board. Carpenter paused to study the board. "Colonel Kim, may I ask where you born?" he asked.

"Seoul," Kim replied. Carpenter made a move and Kim responded.

"Your English is excellent. Where did you learn it?"

"At the University of Seoul and at Berkeley, in your state of California," Kim said, now looking annoyed. The men traded moves a couple times more. Kim started to lose focus on the game with

"I have never been to California, I hear it is nice. I hear that the Capital Division is called the "Tiger" division."

"It is the Brave Tiger Division," Kim corrected.

"That is right! How could I forget that? How long have you been married?" Carpenter then moved a piece and Kim suddenly glared at him. "Just trying to be friendly, Colonel."

Kim turned back to the game and Carpenter gave a faint smile. A few moves later, Kim leaned back and reflectively looked at Carpenter. The Korean Colonel shook his head negatively. He reached for his king, paused, then laid it on its side. A gasp erupted around the bunker from the ROK troops as well as the American soldiers. Kim partially stood, then reached across the chessboard and shook Carpenter's hand. Carpenter grinned widely. Instead of leaving the table, Kim sat back down. "Do you have time for another?"

"Of course, Sir!"

"Good, set up the board and I will be right back." Kim got up and headed towards his quarters. Whitehorse and Ransom stood up and patted Carpenter on the back. Carpenter waived them off so he could concentrate on the next game. He set up the board and waited for the officer to return. Carpenter was white for this game.

Ransom whistled a low note. "Carpenter beat himself an officer. Do you think you can do it again this game?"

"I think so."

"This will definitely be interesting," added Whitehorse.

A few minutes later, Colonel Kim came back and sat down. There was silence that fell over the bunker complex as many more individuals from both armies were now very interested in the contest as well. Whitehorse looked around and there must have been twenty-five soldiers, Korean and American, now circling the table to watch the mental combat. Quiet side bets were exchanged among the troops. Both side's pride was on the line.

Carpenter started the game with a standard opening and attack and was blocked by Kim. The game quickly progressed but after multiple moves, was basically a draw. Both men were lost in acute concentration, tuning out the soldiers around them who watched intently. Kim rubbed his eye, as a signal.

Suddenly, Sergeant Park walked up to the table and whispered in Kim's ear.

"Standard Orders Sgt Park. The officer on duty determines if the man found is a deserter. Then the deserter is dealt with. Shoot him," Kim replies. "Those are my standing orders." Carpenter is visibly startled. Kim made a move and then commented, "Don't you agree, Corporal Carpenter?"

"Agree to what, Colonel?"

"That those who abandon their brothers in arms, are deserters, and should of course be shot?" he asks coldly, staring Carpenter in the face. Suddenly the Colonel had become very serious.

"I think...well...I...that is not my place, Colonel," Carpenter replied meekly. He made his move on the chessboard. He was not comfortable with where the conversation was going.

"Oh, come on, Corporal, for good order and discipline, examples have to be made, surely you must know this? I believe the U.S. Army has the same policy?" Kim then made his move and leaned back in his chair, smiling. The room had grown deathly silent. All of the soldiers now looked at Carpenter, waiting on his response. "Don't you agree, Carpenter?"

Carpenter made a move in response. He seemed agitated, unfocused. "It depends on the situation," he said finally.

"That is what a leader decides. The situation, then he follows orders for good discipline and order. Don't you agree?" Carpenter now looked visibly shaken. He looked around, as if looking for someone to help him with the answer. Not finding any, he made his next move.

"Based on the situation and circumstances, the priority would be for good discipline and order," Carpenter finally said. Kim smiled and moves his piece.

"I am glad you agree, Corporal. When you rise in rank, you will understand this more clearly."

Carpenter looked at Kim, then looked at the board. His eyes

went wide and he realized he'd being beaten. Kim looked smug. After a few more moves, Kim said quietly, "Check mate."

The Americans groaned and the ROK troops in the bunker cheered happily. The two men shook hands over the board and then both stood.

"Thank you for the game, Sir," Carpenter said firmly.

"You are welcome. It was diverting." Kim left the room and a path was made for him. The noise level in the bunker went up dramatically as bets were settled.

Ransom walked up to Carpenter, "What happened?"

"Our buddy got distracted and lost his focus," Whitehorse replied. He then held up a fistful of cigars and smiles. "Our winnings!" Carpenter looked at him and shook his head.

"You bet against me? Again?"

"He did unto you what you did to him in the first game. And by the way, no one was shot." The three Americans moved away as another two men sat down to play. The show was over.

Carpenter and Whitehorse walked into their shared quarters. Whitehorse squatted down to a small makeshift stove made of ration cans. The stove was vented into the kitchen chimney. A coffee pot sat on top. Whitehorse grabbed a rag and poured the hot liquid into two U.S. Navy coffee cups that were hanging on the wall nearby.

"The stove you made works well," said Carpenter. He gratefully sniffed the aroma of the coffee and smiled. "Please thank Lt Dak for the coffee and the mugs. That guy can get anything for the right price."

"And thank you for my winnings!" Whitehorse responded.

"Ha!" He smiled and sipped the hot brew. Lt Col Kim then walked by their quarters, with Lieutenant Dak, on their way deeper into the bunker. Carpenter stood and followed. He noticed the air of respect quietly given to the Korean battalion commander by troops on both sides. The lieutenant and the colonel headed down to the area of the Buddhist shrine. Once they arrived, they lit some incense and knelt to pray. Dak finished quickly and turned to leave. Kim lingered on his knees. On his way out, Dak saw Carpenter and whispered, "He prays for his family, his men and his ancestors. His family has been missing since we retreated last year from Kumsong. He has lost many brave men and prays for their souls. He respects his ancestors' customs and provides an example for all of us."

Carpenter quietly nods his head in acknowledgement. He looked at Kim one more time, then turned and walked back to the main common area, leaving the colonel to his devotions.

The situation was confused for a long time, but we thought we were pretty secure, as we were still pretty much behind the front lines. In mid-July all hell broke loose. The Chinese flooded into the area, and after two days of heavy fighting the front lines seemed to collapse. Korean troops were rolled back through our area in a rout. Many of them had thrown their weapons and gear away. It was at this time that Lieutenant Colonel Kim arrived and took command of the complex. He was a Battalion Commander, whose Battalion had disintegrated under the Chinese Assault. He was mad as hell, but well organized. He immediately sent some of our Guard Troops out to form a line along the ridge,. As Korean Soldiers came back to the line, they were allowed through if they had no weapons. However every soldier who had a weapon was directed to report to our complex. In just a few short hours we nearly doubled our garrison, and gained five BAR's.

Richard Carpenter, Memoirs

Chapter Seven

Several days later, above ground, a jeep roared up to the bunker complex. Three Americans were sitting inside. An officer got out and headed to the command bunker. The driver remained in the driver's seat but the other, older American got out of the jeep and sniffed the air. He then followed the aroma to the vent pipe which confirmed his suspicions. The man then went down the ramp into the bunker and found the source of the pleasant smell. The driver, suddenly understanding what the other American had found, followed quickly down into the bunker area.

"What does an air force guy have to do for a good cup of joe?" the man yelled into the room where Whitehorse and Carpenter were sitting, enjoying a morning coffee.

"Just come on in and tell us what the Hell is going on out there that we don't know!" Carpenter replied.

"I'm Sgt Frazee, and behind me breathing loudly is Airman Cushing. They call me 'Blue Fin'. Our Lieutenant is upstairs," said the first American.

Without any further comment, Frazee barged in and grabbed a mug, then filled it with coffee. He then handed the pot to Cushing who then poured himself one. Carpenter handed over an Army ration of sugar. Cushing nodded his head in thanks. Frazee shook his head no and settled on Sgt Park's bunk, smiling.

"Where's Park?" Frazee asked.

"He's up near the front with Colonel Kim," Whitehorse responded.

"Blue fin? What the heck is a bluefin?" Carpenter asked.

"What, you don't know? Where are you from? Some farm in Iowa? It's a tuna. Now you know."

"He's from North Dakota," Whitehorse added. Frazee smiled

and looked at Carpenter in understanding.

"Ah! Betchya ya was a farmer. And you never fished in the ocean, didya son?" Carpenter shook his head no; he had not figured out Frazee yet. "The best fishin is ocean fishin. Get a tuna on a hook, ya don't even need bait when they are running. They have the best white flesh that is tender and juicy…"

Suddenly a Korean shout came from the next room. The men rushed through the door to see a ROK private being confronted near the ROK NCO bunk area. Two Korean NCOs held the man while another searched him. They found a knife, a watch, and some other items.

"What?" Carpenter said.

"Thief," replied Whitehorse.

"You betcha," said Frazee.

The Korean NCOs asked the man angry questions. When they got their answer, they went quiet as the realization hit of what the man had done and what the consequences would be. They paraded him out of the room to the open area outside.

Lt Dak walked toward the Americans looking ashamed. He stopped in front of Carpenter and gave an apologetic bow. He handed Carpenter the pocket knife. Carpenter stared blankly for a moment then started to pat his pockets. When he realized his knife was gone, he took it from Dak in silence.

"My apologizes. It will not happen again." Dak then walked to the doorway of where most Americans lived.

"I must return this item. Who owns this watch?" Brenner stepped forward and got his watch. "My apologizes. It will not happen again." Dak was stiff lipped and he turned away and headed up the ramp.

Frazee and his group went back inside the room then commented, "Well that ROK won't be around for the big show coming."

"When is the big show?" asked Whitehorse.

"Don't know. Maybe in a week, maybe more. No doubt about it tho, it is a comin. The signs are there boys, the signs are there," replied

70

Frazee. He looked around as if he was missing something. "Got any of those little coffee cakes in a can?"

Carpenter quit looking at his pocket knife and dug in his bag and tossed over a can. Frazee used his little army can opener to open it and smiled his thanks. "Too bad ya guys won't be around for the fireworks now that your time is about up!" he added.

Carpenter looked up startled, "How do you know that?"

Frazee laughed, "Ya'll are short timers. Word gets around up here. Any news is good news. Some say you are testing a death beam, others communications, and I even heard you are training. Nope, my money is on infrared or radar. That is why you do the testing at night."

Whitehorse responded, "Maybe yes, and maybe no." A distant pistol shot echoed through the bunker. Carpenter and Whitehorse flinched, while Frazee was unfazed by the execution of justice. Carpenter looked unsettled. Whitehorse looked at Carpenter questioningly.

"Hopefully, when ya'll leave, ya'll leave me some of that fine coffee?"

Our complex bristled with machine guns. Six of the eight 50 Cal. guns had tripods, and were modified for ground mount firing. The remaining two did not have tripods, and we used them as spares. Each of the half tracks also had two ground mount 30 Cal. machine guns. So that added another eight guns to our lines. Plus the Korean guard unit had four 30 Cal.. Machine guns, two 3.2 rocket launchers, and three 60 mm mortars. We were able to construct a significant defensive perimeter. With nearly four hundred men, we nearly stood shoulder to shoulder in our positions. It did not take long, and we could see the Chinese flooding across the ridge line near the bunker to the south of us, and we came under direct attack from the south west. The attack lasted for about four hours, and we could see the pile up of bodies, as one wave of Chinese after another tried to reach our positions. Because of our fire power we held.

That same night we set up our radar units and immediately

identified a buildup of Chinese troops getting ready to attack us straight down the ridge line. We opened fire on them while they were still being assembled, and the attack fizzled out almost as soon as it began.

Richard Carpenter, Memoirs

Kimpo Airfield
Seoul, South Korea

The driver of the deuce-and-a-half could see the jeep in his rearview mirror, closing at a fast rate of speed. He subconsciously applied the brake in order to allow the speeding jeep (G-eneral P-urpose) room to pass. As the jeep approached the rear of his vehicle, it careened around the truck's left side, attempting to make it on time to a staff meeting on the other side of the airfield. Major Hastings, the officer in charge of Lieutenant Sharp's detachment's mission at the U.S. Army Security Agency headquarters in Seoul, was riding in the passenger's seat. As the driver of the jeep veered back into the right lane, an elderly, Korean papa-san attempted to cross the roadway with a heavy load on an A-frame, not seeing, or hearing, the speeding jeep. The jeep's driver swerved around the old man, caught a tire on the edge of the road, and lost control. The jeep landed in the ditch to the right side of the roadway. Major Hastings was thrown clear. The driver, having stopped the vehicle, ran back to find the major lying by the side of the road, blood gushed from his head, nose and mouth. The driver of the two-ton truck had also stopped and ran to the scene. "Is he dead?" he asked.

"I don't know if he will make it," said the jeep's driver.

Hill 433
Officer's Quarters

The sounds of war were getting louder and the command staff of the bunker were getting nervous. They knew the hilltop position could not hold against an unrelenting Chinese assault forever. Lieutenant Sharp finished writing, sealed an envelope addressed to Major Hastings and handed it to Sergeant Quincy. "Sgt Quincy when you get to the ROK Capital Division HQ, stress that this envelope is very important to the KMAG officer. It must get to Kimpo with all dispatch. Hint, if you need to, that our testing time is up and it is time for us to leave."

"Yes, Sir! I wholeheartedly agree!" Quincy saluted and left the bunker on his way back to Kimpo near the South Korean capital.

ROK Capital Division HQ
IX Corps Sector
Korean Military Assistance Group Officer's Desk
June 1953

Sergeant Quincy walked up to the officer behind the desk and saluted. "Sir. Per my orders, this envelope must get to Major Hastings at the ASA compound at Kimpo as soon as possible."

"What is the rush, Sergeant?" the officer asked.

"Sir. I am not at liberty to say. It is Top Secret. And it is important, sir. We are finished testing and must ..."

"Okay, okay, I get the idea. Let me see the envelope." The

officer took the envelope from Quincy and frowned. "It is addressed like a letter. Don't you have classified, carrier envelopes?"

"No, Sir. I was hoping you could help me out with that, Sir."

"I will!" The KMAG officer took out a secure courier bag and wrote a note that read, *American AA unit on Hill 433 requests relief as per orders. Please advise. ASAP.* He wrote *Urgent* and addressed the courier bag, placed the envelope inside and locked it. "Sgt Quincy, it will go out in less than an hour. Anything else I can help you with?"

"Yes, Sir. Our American rations have been stopped. Can you renew them until we are relieved?"

"I will look into it. Anything else?"

"No, Sir. Thank you, Sir!" Sergeant Quincy saluted and left.

U.S. Army Security Agency Compound
Kimpo Airfield, South Korea
June 1953

The Army security agency courier walked into the ASA HQ and handed the classified document carrier to the duty noncommissioned officer, who immediately signed for it. The courier, his day's mission accomplished, left the premises. The NCO opened the bag and read the letter. He immediately took the letter to the duty officer.

"Hill 433? An AA unit asking us to be relieved? That is strange. Do you know anything about them?" the officer asked.

"No sir. It was addressed to us and all the bag had in it was a letter to Major Hastings and that note."

"Major Hastings is in Tokyo. He is seeing a specialist due to his accident. He will be back in a few weeks. Don't forward it. Just put it in his in-box."

"Yes, sir. And the response?"

The duty officer writes, "Hold position until further notice." He signed it and handed it back to the NCO. "That should hold them for now. I will ask around about this AA unit on Hill 433." The NCO left the room with his instructions. The officer picked up the phone to close the loop on the issue.

"Hey George. Do you know anything about a special AA detachment on loan to the ROKs up with the Capital Division in IX Corps? Hill 433? No? Ask around for me, will you? Let me know if you find out anything." He paused, and added "Thanks," and hung up the phone. The officer then went about his other tasks.

Chapter Eight

Hill 433

The noise was deafening as the battle commenced. Thousands of Chinese had massed and attacked ROK lines ahead of the detachment's position. Artillery pounded both sides and machine gun fire raked the Chinese and Korean positions. The forward air control team was actively calling in artillery and air support against communist lines. The FAC officer yelled, "Frazee, right 100, fire for effect!"

Frazee responded, "Roger, right 100, fire for effect," He grabbed the radio microphone, "Fire correction! Blue Fin Actual wants right 100! Pour it on em!"

The ROK troops continued to put up a fight, earning the respect of the Americans. "Good God," said Frazee as he sees the Chinese troops approaching. "There are thousands of them," he muttered to himself.

The next day was quiet. We could see Chinese troops crossing the ridge line about a mile south of our position, but none in our immediate area. But we knew that it would be only a short reprieve. That night we set up our radar again, and immediately detected Chinese troops assembling, and lining up for an attack on the slope to our north west. Again we employed our 50 Cal. machine guns to give grazing fire across the slope, and into their positions. Again the attack fizzled. This time they simply ran for cover.

Richard Carpenter, Memoirs

Bunker, Hill 433

Carpenter laid down his hand of cards, "Read 'em and weep boys!"

"Crap! I should have known!" said McKenzie. He threw down his cards on the table, pissed off at himself for losing. Carpenter smiled and collected the cards.

"Better luck next time," added Carpenter. "My watch starts in five minutes! You'd better get out of here!" he said, laughing.

"Yeah, yeah. Shit! Shit! I should have known! Damn it!" McKenzie stomped out while the rest of the men were laughing. He walked out into the trench, looked up and saw the stars. He followed the trench line forward past a few ROK soldiers that were on duty. As he approached the forward bunker, he called out, a little too loudly, "Hey Lee, I'm coming in!" Two ROK soldiers checked his face and identified him, then allowed him inside. It was even darker inside than outside.

"Where's Carpenter?" asked Lee. "He was supposed to relieve me?"

"I just lost a stupid bet to him. I should have known better!" Lee laughed and handed McKenzie a flashlight. He turned it on but the light was dim. He smacked it against his hand and it burned brighter for a minute.

"Yeah, you should have known. I lost a bet to him a few days ago. He wins more than he loses. I hate playing with him. He got me to do his shift last week."

McKenzie groaned and then smacked the flashlight as the beam grew dim. He was counting the ready grenades and was not being careful. He moved them around, repositioning them, and then the flashlight died.

"Shit, the flashlight is dead. Where are the batteries?"

Suddenly the men heard a hissing sound. A look of horror washed over their faces in the dark. There was no need to see each other, they knew what had happening.

The ROK soldier screamed in Korean, "Live grenade! Put it in the grenade hole!"

The other ROK soldier added, also in Korean, "Hurry up! Hurry up!"

Lee also shouted, "Grab that live grenade and dump it in the hole!"

McKenzie beat the flashlight as he searched for the live grenade. He started throwing every grenade he found into the grenade hole.

"Drop it in the hole! Get it! In the hole…" Lee shouted.

The soldiers along the trench saw a flash of light and heard the explosion. Smoke poured from the bunker.

As the smoke cleared, Carpenter came running down the trench. He heard the cry of someone in pain. He dove into the bunker entrance. In the smoky darkness he tripped over a body. Using his hands he felt upward toward the face. He shook his head. There was only bloody mulch where the face should have been. He peered around coughing in the smoke and stumbled over another body. It too was unrecognizable in the darkness. The cries had now become whimpers. Carpenter moved toward the sound and found two dead ROKs. He also found the wounded ROK soldier. He coughed in the smoke and started to carry out the wounded ROK soldier.

Kowalski arrived and saw that the first ROKs to arrive were pointing their weapons into the darkness, looking for the enemy. He nodded his head, then heard someone coming out of the bunker. He watched Carpenter stagger out with his load. He had blood all over his hands, arms and covering part of his now very pale face.

The fourth ROK soldier cried, "Medic! Medic!"

"What the Hell happened here?" shouted Kowalski.

Carpenter relinquished his burden to a ROK aid man. He coughed. His face showed anguish and he looked distraught. He tried to take in a deeper breath. He coughed again… "McKenzie & Lee. A grenade went off in their face…"

Carpenter then retched. He turned away as he threw up and

gagged until his stomach was empty. The Americans had begun to die.

Later that evening, the Americans were operating one of the radar sets near track 4. Night had fallen. "Hey, Da-vid, rotate it back to the right. Hold it right there. Target!" said Unger.

Da-vid came down from his sandbag perch and looked at the screen while Unger used the power phone to contact the command bunker. "The targets are on the road. It must be a ROK patrol," David responded.

"This is track 4. Radar reports multiple targets coming up the road. It looks like a ROK patrol." There was a brief pause. "Yes, sir. We picked them up on the road as they came around the bend. They are now about 250 yards out." Another pause. "Yes, sir. Going to alert status!"

Da-vid watched the radar monitor and listened to Unger. He then turned toward a ROK soldier looking expectantly at the Americans. He smiled and told the ROK... "Alert! Alert! Alert!" The ROK soldiers sprinted off to spread the word.

Carpenter and Whitehorse were off duty, sitting looking up at the stars in the sky, when PFC Earnest Spivey came rushing down the trench from the command bunker. "ROK patrol coming up the road from the valley. It is not one of ours!"

Spivey rushed down the trench line. Then came Sgt. Park with SSG Kowalski. Without a word they too headed off toward the east side of the bunker complex. Trailing a bit was Lt. Sharp, who was adjusting his web belt with his 45 caliber pistol in its holster. He saw Carpenter & Whitehorse.

"We got visitors. ROK patrol."

Whitehorse looked up the trench toward the command bunker as if expecting another person to be coming. As he turns to Carpenter he commented, "Parade over?"

Carpenter laughed and dropped down into the trench, then

mimicking Sharp, adjusted his web belt and pistol holster. "We got visitors." Carpenter then headed east down the trench toward the guard post. Whitehorse followed, adjusting his web belt and holster as well.

Three ROK soldiers were manning the forward outpost when they saw the unidentified patrol walking down the road.

"Halt!" they ordered. Then in Korean, "Password!"

The officer leading the patrol responded, "Patrol halt! I am Lt. Han, 6th Division. I have important dispatches for the Regimental Commander of the 26th Regiment."

"Move five paces forward and stop!" ordered the ROK guard in Korean. The patrol leader moved five paces forward and then a high-powered flashlight blinded him.

He cursed in Korean and added a common South Korean curse, "You messed up my night vision!"

"Yes, Sir! Standard procedure."

"You did right soldier. I could have been an infiltrator. Now use your landline to let them know I am here with dispatches. I have 10 men including me." One of the ROK soldiers picked up the power phone and cranked the handle. Another was covering the patrol with a BAR.

He waited a second then reported, "This is outpost 1. A Lt. Han from the 6th is here with dispatches and with 9 man patrol." He waited for a response. "Yes, Sergeant, I will send them up." He then signaled the okay visual code, the thumb and fore finger flashed twice to the ROK soldier in charge.

"Lt Han, you are cleared to the wire. Stop there."

"Very good. Patrol, forward march!"

The ROK soldier watched the officer head up the road followed by another soldier. The second soldier had on a backpack. The soldier looked straight at the ROK guard and nodded. There was enough light to see he was Korean and not Chinese. The other soldiers walked slumped down, shoulders appearing tired, with their helmets down. The

ROK guard counted the members of the patrol and confirmed ten men total. He turned toward the other soldiers of his outpost. "Confirmed ten men. That NCO looked like a mean one!" The guards then turned their attention back to watching the road.

The patrol walked further towards the bunker area and approached another ring of ROK troops guarding the perimeter. The South Korean soldiers were wary. The men challenged the approaching patrol. "Halt! Identify yourself!"

"Patrol halt. I am Lt. Han, 6th Division. I am expected."

"Step forward 5 paces and halt!" The officer followed orders and once again a flashlight beam lit up his face. He flinched.

"You enjoy that?"

"No, sir. Standard procedure."

"Well, you're lying. I enjoyed it when I was an enlisted soldier. I guess it is payment for what I used to do!"

ROK Senior Sgt Park was watching the other ROK NCO go through the appropriate motions. He nodded his head, then heard the officer's comments and chuckled. He saw another soldier with a packback step forward.

"This is my Sergeant. He carries the dispatches. May my men rest here while I conduct my business? And which way to the command bunker?"

"One minute, sir."

Park watched the soldier check with the Command bunker NCO in charge. There was no warning signal given. He relaxed somewhat.

"Yes, sir. Down the trench to the to the second right. Then straight ahead past the underground bunker and you will walk into the command bunker."

Park watched the exchange and checked out the two men as they walked by. The light shifted brighter, due to a break in the clouds and he saw their eyes and verified they were Korean. The officer nodded at him. Park then turned toward the members of the patrol who were moving forward instead of sitting down where they were. He looked puzzled and stepped forward. One of the visiting soldiers stepped up on the sand bagged edge. Park's eyes went wide. Quickly, he pulled his 45 and aims it at the intruder who was wearing Chinese sneakers. He fired at center of mass then yells, "Infiltrators! Stop them!"

Gunfire erupted as the ROKs responded and shot at the visitors. Park suddenly turned and saw the infiltrator Lieutenant shoot down the American Sgt. Kowalski. The American Lieutenant turned and ran back down the trench. Behind the fake Lt. was the infiltrator with the backpack. His disguise now blown, he turned to allow his North Korean Comrade time to blow up the bunker. He saw the ROK NCO (Park) who he just passed a few seconds prior now aiming at him. They began to trade pistol fire. PFC Spivey saw Sgt Kowalski go down and Lt Sharp turn and run toward him. He plastered himself on the side of the trench to make room for Sharp to pass. He was still there when the soldier with a packback ran towards him. Spivey stayed put, frozen. He heard the shouting and shooting. He stuck out his foot. The enemy with the backpack tripped and went face down onto the duck board at the bottom of the trench. The man grabbed for his pistol and without aiming fired it over his shoulder. He scrambled to his feet to continue. Spivey avoided the gunshots and football tackled the enemy. They both went down as Spivey yelled, "Stop!"

Suddenly a gun went off near Spivey's face and he jerked with the impact. Spivey grimaced and he tucked his head below the backpack. He heard others rushing toward them. Suddenly the North Korean relaxed and there was a blinding white explosion as the backpack detonated. The infiltrator and Spivey were instantly killed.

Sgt Park saw the infiltrator officer run out of ammunition and start to reload. An explosion was heard and smoke roiled down the trench toward them. Sgt Park charged, firing his last round which hit the enemy in the gun arm. Park drew his combat knife and jumped on him plunging the knife down several times. The infiltrator was killed.

Carpenter and Whitehorse, hearing the yells and shots, now drew their weapons and were shooting at anyone above ground in front of them. They almost shot Sharp as he rushed down the trench toward them. A large explosion rocked the trench line and a cloud of smoke followed by a shock wave roiled the trench. Then unexpectedly, there were no targets left and a sudden silence. They reloaded. Lt. Sharp rallied himself and joined them moving down the trench with Carpenter in the lead. They passed a blown out cratered section of trench, but no whole bodies were seen.

Sgt Park reloaded then called out to the other ROK NCO. The man replied then started to call out to his men. He moved forward and checked on the American Sgt Kowalski. He was still alive.

Park yelled in English, "Sgt Kowalski has been shot! I need a medic!"

"Sgt. Park! It's me! Carpenter! I am coming up the trench. Don't shoot!"

"Stay low. Help him. I must do my duty."

Park waited until Carpenter was with Kowalski. He could see other men behind Carpenter in the trench. Park then turned to assist his men in the cleanup. Carpenter, nodded and attempted to stop the bleeding, telling Kowalski that everything will be okay. "Hold on Kowalski. Hold on! Whitehorse get the medic

Now!"

Whitehorse turned and ran back down the trench to get the medic. Lt. Sharp with his pistol out, stared at Kowalski.

"Don't just stand there. Help me stop the bleeding!"

Sharp put his unused pistol away. He grabbed his bandage from his web belt and applied it where Carpenter directed. Suddenly the medic (Casper) appeared and they worked frantically to help Kowalski.

Later, Kowalski opened his eyes and looked around. He saw the medic and Carpenter and a ROK medic working on him. "I got hit? How bad?

"Bad enough. Are you in pain?" Casper responded. Kowalski rolled his head negatively.

"Nah. Not really. I am tired."

"Hold on Sgt. We need you, so you just hold on."

Kowalski smiled then lolled his head from Carpenter and saw the worried face of Lt. Sharp. He reached out toward the officer, and Sharp moved forward. Kowalski grabbed his arm and with great effort said, "Take care...ah, I need to, ah....I need..." Kowalski passed out then died a few minutes later.

As the medic confirmed the death, there was gret sorrow on Carpenter's face. In the background, Lt Sharp said, "Carpenter. You take over Kowalski's squad. You are a Sergeant now."

Chapter Nine

About midday a Chinese loud speaker started to tell us to surrender. It was being broadcast from a position under the cliff to our north east. It annoyed us a little, but we did enjoy the music they played. That night one of our sergeants and a couple of men went out to the top of the cliff and set up a large 50 gallon barrel full of gasoline, and set an explosive charge under it. The next time the loud speakers came on, we fired off the charge and blew flaming gasoline over the cliff. The speakers went silent, and we never did hear them again.

Richard Carpenter, Memoirs

The next morning brought sadness to the bunker complex. The sun rose to multiple body bags lying near a two-ton truck that was to carry them to the rear. They included the dead ROK soldiers and two that were set slightly aside with Carpenter, McCoy, Whitehorse and Ransom standing over them. A little farther away was a pile of dead infiltrators with their ROK uniforms stripped off them, being loaded into the back of another truck. Park was watching them.

Carpenter and Whitehorse picked up Sgt Kowalski's body and loaded it on the truck with the other ROK dead and the few ROK wounded being sent to the rear. A wounded ROK soldier helped from the truck. Then Ransom & Whitehorse loaded the broken remains of Spivey into the truck. Carpenter pointed at Spivey's remains and emotionally stated, "This man stopped the suicide bomber. His name was PFC Ernest Spivey. He is, was, a good soldier." He then pointed at Kowalski. "That man was my Sergeant. His...his name was Sgt Mike Kowalski. I will miss him." With tears in his eyes, Carpenter turned away. The wounded ROK soldier nodded solemnly and sat down. He looked at the covered American bodies and the tail gate went up as the truck pulled out. Carpenter watched the two trucks disappear into the rising sun. Then he turned and walked back to the bunker compound. There was still work to be done.

The NCOs and Lt Sharp stood in the command bunker. The sound of artillery thundered in the distance. Machine gun fire peppered the canyons, creating a deadly cacophony of sound. The war was far from over, as all of the men had initially hoped.

"I've got good news and bad news," said the Lieutenant. The NCOS breathed a sigh of disappointment that was palpable in the small room. "The good news is that the ROKs seem to be holding under heavy pressure from the enemy. The bad news is that we have been told to, quote Hold position until further notice unquote." Sharp passed around the request to be relieved and the order to stay put. He bit his lower lip.

"The front line keeps coming closer and closer. We have finished our testing. We should have been out of here some time ago. We may have been lost in the paperwork. We are in limbo. I am sure they will get us out of here. Let your men know I am working on it. Sgt Quincy. Would you be so kind as to brief McCoy and Carpenter on the situation to our front?"

"Yes, sir." Quincy waited for a moment then started talking pointing at a map. "To our west nor-west across the valley is Hill 536. That is old Sniper Ridge. It is held by the ROK 9th Division. Just across the Namdae ch'on is Hill 424 held by most of the 1st Battalion of the ROK 26th Regiment. The valley not only has the river but a railroad line to the north. The road is 117 and the adjacent rail line goes to Kumsong. To our northeast is Hill 453 that is where Lt. Col. Kim is at with the 2nd Battalion of the ROK 26th. That is why we have only a few ROKs with us right now. Hill 453 is his forward command headquarters. To his north are Hills 477, 522 and 512. Hill 512 is held by the first battalion of the ROK 2nd Cavalry as is Hill 552 to their east. The enemy has now taken Hill 467. And Hill 468 next to the northern portion of Road 117A is in danger of being overrun. Reports from the 6th ROK indicate they are moving units to support Hill 468."

Quincy paused and McCoy commented, "It looks like a hockey stick with with a bent handle."

"Where are our artillery units?" Carpenter asked.

"Most of them are along Road 117A from 117B past 119 up to Road 120. They are mostly road bound because of the terrain. Oh, one last thing. The Capital Division headquarters is in Chaegung-dong with the Cavalry reserve. This attack appears to be the start of the Chinese Summer offensive."

"So much for the peace talks," added McCoy.

Sharp took over the conversation, "I don't like it, but we have our orders. We must stay here. Sgt Quincy, how are we set for ammo, rations and water?"

"We have plenty of ammo and water. Our American food is about gone. But..., we have plenty of rice, spam and dried fish."

The men laughed politely. Then the command bunker phone rang. The ROK operator talked for a moment then turned toward Lt. Sharp holding the phone out to him. Sharp answered. "Lt. Sharp here. They are on Road 117 pushing down 117B? Okay." He paused. "Yes, we can do that. I'll get my men ready. Yes, sir.

I'll call you when we are ready. Out."

Sharp handed the power phone back to the ROK operator. "That was Lt. North with the FAC team. The enemy has broken through 1st battalion's lines in the valley. The FAC wants us to use what we have to interdict the eastern ridge line to our north. After we and the fly boys do their thing, Lt. Col Kim will counter attack west from Hill 453. I want every one up and ready including the radar. Move it gentlemen! It is time to fight!"

McCoy and Carpenter raced out to rouse their squads to full alert. Quincy headed down stairs with a ROK soldier to sound the alarm there.

Carpenter watched Lipscome and McKenzie working their men. He watched as they reported ready and received their targets. Their 40mm started barking and the loaders added 5 round clips to the top of the ammo guide. He then heard the other 40 mm start to fire. He turned his binoculars on the target and saw the 40mm rounds explode on impact and those rounds missing going out about 4500 yards before air bursting. He saw ants being tossed about. Carpenter paused and noticed that his Quad 50 was not firing. He slithered down and asked Tucker what is going on.

"No targets for the 50s?" Carpenter asked.

"Nope. Most of the enemy are just on the other side of the ridge. We can't indirect fire."

"Bullshit! The 50s can drop rounds indirect," Carpenter responded.

"No fuckin way!"

"What goes up must come down. Right?"

"Five bucks says you can't hit where they want you to."

"Only $5? How about more on that?"

"Sorry. I haven't been paid in a while."

"Okay. That is a bet. Let me have that phone." He paused as the radio operator handed him the receiver. "This is Carpenter. I would like to try to walk in the 50s from track one as indirect fire...No I am not kidding, Sgt Quincy. But I will need the FAC team to see where the rounds hit...You can set that up? Good. I'll let you know when we are ready to fire. Out."

Carpenter tossed the phone to Tucker and laughing, ran down the trench to Track 1. Just before he got there, he started yelling. "Whitehorse! You got a fire mission!"

"What? A fire mission? At what?"

"The other side of the ridge!"

"No way. We can't see there. Can we really do that?" added Ransom.

"Hell yes, and we will." Carpenter used his hands to show what he wanted. We fire upward at 45 degrees elevation. The FAC team will advise where the rounds hit. Then we will adjust."

"How will they see it? We only fire one in six tracers?" asked Whitehorse.

"We have some old belts of one in three. We didn't break them down yet," said Ransom.

"Remind me to chew you out later! Load them up!" Whitehorse added.

Carpenter watched the Americans and ROKs load up the special links and elevated the Quad 50s to 45 degrees. Quickly they were ready.

"Carpenter! I got the FAC team on the phone. It's your buddy Frazee," yelled Sam Smith.

Carpenter rushed to the phone. "Frazee! We will fire one in three tracers at 45 degrees. This will impact the rounds somewhere down slope of the ridge. Ready? Good. Hold one!" He turned to track 1. "Commence firing!"

The Quad 50 shuddered as 4 50 caliber machine guns fired into the sky. Six hundred rounds went flying out, then a pause. The loaders linked the next belts.

"Great Balls of fire!" Whitehorse exclaimed.

"Outgoing!" Carpenter yelled into the phone. All eyes were on Carpenter as he waited for results. "One hundred fifty yards too far and right 75 yards!"

"That is up, ah 50 or uh…" said Whitehorse.

"Up 52 degrees and right 5 degrees. Ready to fire," said Ransom.

"Fire!" ordered Carpenter.

Again 600 rounds flew out in the heavens, every third round a tracer. Carpenter turned to the phone when the gunner paused and yelled, "Outgoing!"

Another pause as all but the loaders watched Carpenter. He yells into the phone, "What? You're kidding? … Great! Roger that! Fire for effect! Stand by."

Carpenter turns and yells at Track 1, "Fire for effect! Then walk it left a degree at a time."

For several hours the battle raged. Lt. Col Kim counter attacked and for the moment, the danger was past. Then during a lull, some ROK trucks rushed forward with reinforcements and they come back with wounded and dead ROKs. Smoke billowed and machine gun bursts rang out from a distance. Carpenter was talking to Whitehorse and Ransom while they watched the ROKs service the quad 50s.

"Sharp said what?" asked Whitehorse.

"He said our orders are to stay put. I saw his request and the order to remain here."

"Shit!"

"Not only are we a bastard outfit, now we're a lost bastard outfit!"

"Lost Bastards, that sounds about right!"

The stalemate of the Korean conflict lasted for two years, from 1951 to the armistice which was eventually signed in July of 1953. Little territory was exchanged during the last two years of fighting and the war became a test of wills. The Chinese were suffering tremendous casualties and endured major logistical problems. However, they hoped to grind down the Americans until they no longer had the stomach for the fight. One of the major issues to be negotiated was the return of prisoners of war. North Korean and Chinese prisoners did not want to return to the communist North. This was unacceptable to the Mao and the North Korean leadership. The South Korean goal was to recapture the original territory of the South and to prevent losing any territory. The North Koreans and the Chinese also used psychological operations to test the American and South Korean will to continue the conflict. In June of 1953, the negotiations drug on while the soldiers on both sides fighting for Hill 433 continued to die.

In a night artillery and human wave attack, the ridge line

previously held was taken by the Chinese Army. The ROKs retreated back a few hundred yards. There they rallied and waited for daylight as their own artillery hit their former position.

As day broke, Carpenter watched the shadow retreat from the ridge line in front of Hill 433 with his binoculars. The telephone rang in the morning quiet. He looked down and stuffed cotton in one of his ears. Then he paused as he waited for Rodriguez.

"Stand by to fire!" Rodriguez yelled.

Carpenter looked over and saw the Quad 50 ready. He turned and saw the 40 mm position ready. He stuffed the remaining piece of cotton in his other ear as Rodriguez yelled, "Commence Firing!"

Only the first word was heard as the quad 50 started spraying the ridge line from left to right. Carpenter watched the 1 in 6 tracer rounds fly out and impact near the top of the ridge. Occasionally, a tracer hit something solid and the round bounced, usually upwards. Carpenter saw the 40mm 1 in 5 tracer rounds fly out and explode on impact. Even with the binoculars he saw humans the size of ants being tossed by the fire power being thrown out. Carpenter then scanned the lower areas still in darkness. Here and there he saw the ROKs attacking up the hill. He spotted a machine gun nest and decided the 40 mm would be a better weapon. He slid down the sandbags and ran to the 40mm position. As he got there, he tapped Smittie on the shoulder and motioned for headphones and microphone. He got it and moved far forward to re-sight the enemy machine gun nest. Into the microphone he yelled, "Gunner hold fire! New target!" He then paused for the gun to be quiet.

"Left 10 up 4. Fire two rounds when ready." He paused and observed rounds short & right.

"Left 3 up 2. Fire two rounds when ready." He again paused and observed.

"Left 1 and fire 10 rounds!" Tens rounds flew out and exploded destroying the machine gun nest. Carpenter saw the enemy retreating and the ROKs closing.

"ROK troops are closing in! Rake the skyline. Fire when ready!"

Carpenter slid back and handed the phones to Smittie. He grabbed the other power phone and rang it.

He heard the following on the phone, "Track 1. Rodriguez."

"This is Carpenter. Have the 50s rake the skyline! Try not to hit the ROKs! Got it?"

"Roger that! Out!" Carpenter headed back to his position and watched the ROKs fighting near the ridge line. A sudden movement caught his eye. It was a jeep coming from Hill 453 and dodging mortar round fire. It zigzaged up the hill to Hill 433. Carpenter ran toward the 40mm position and rang the command bunker

"Jeep coming in with wounded! Send medics to the rear! What?" He paused to listen.

"Roger! Cease Fire! Cease Fire!" He paused again. "The medic ... ? Roger. Out."

Carpenter ran toward the rear and got there and saw Airman Cushing driving like a madman coming in hot, honking his horn, with Sgt Frazee holding the passenger in the seat. A bloody bandage covered the chest of the man in the passenger seat. The man's head rolled to one side and he saw it was Lt. North. The jeep stopped and Frazee was screaming, "Medic! Medic!"

Medic Scott Casper rushed forward and examined the man. A ROK medic assisted. Casper turned toward Carpenter declaring, "It is bad. If he does not get to a surgeon at battalion aid soon, he will bleed to death. I can't do anything here for him!"

"Shit. Go. Go!"

Frazee & Carpenter yanked out two portable radios and a gear bag from the back of the jeep. Casper jumped in and slapped another bandage over Lt. North's wound. Cushing took the driver's seat and started the engine and shifted to first gear.

"Get! Get there and come back ASAP! And get more medical supplies!" ordered Carpenter.

The jeep raced off and down the hill. Carpenter turned and only saw the back of Frazee heading toward the command bunker with one of the radios. Carpenter grabbed the radio and the ROK medic the gear bag. Then followed Frazee.

Walking into the command bunker, Frazee started to set up the other radio and patched it into the main antenna line. Carpenter was talking to Lt Dak who ordered the radio and gear bag downstairs.

"Come on. Come on. Power up," Frazee cajoled. He pulled a map out of his jacket and smoothed it out. Then he began fiddling with the radio knobs.

"There. Good. Frequency. Check." He paused, then continued. "This is Blue Fin. Fire mission!" He paused again. "This is Blue Fin 4! Yes." During this time Lt Sharp arrived and was briefed by Carpenter in the background.

"Standing by. Yes! High priority! OP! SOI point 24" Frazee paused again.

"Correct. Roger, watching for smoke." Frazee jumped forward and headed to the spotting scope. He watched the spotting rounds smoke rise. The he rushed back to the radio. "Up 50! Barrage fire. Fire for effect!"

He rushed back to the scope and watched as the artillery rounds came in. "YES! Got those SOBs!" He headed back to the radio. Grinning ear to ear, he picked up the mic and reported, "This is Blue Fin, target destroyed. Good shooting! Blue

Fin 4 out!" Frazee looked up satisfied at Carpenter and Sharp. "Got em!"

At this point, detachment Sgt Quincy came in and confronted Carpenter angrily. "Who the hell gave you the authority to send our only medic back to the rear?" Sharp looked at Carpenter with questioning anger.

Frazee, with blood still visible on his hands and uniform piped in suddenly with an interruption of the conversation, "Lt. Sharp, I really appreciated that Sgt Carpenter allowed your medic to try to save Lt. North by rushing him to the battalion aid station. Lt North got hit trying to see where the CHINCOM set up a forward OP east of 453. If the mortar and artillery has stopped then we just clobbered their OP."

Lt Sharp, divided over the issue, looked at Carpenter, then at the blood on Frazee. He turned to Carpenter, "I'll give you the benefit of the doubt this time. I hope Lt North makes it. And the medic better come back. In the meantime check on your squad. Dismissed Carpenter."

"Yes, Sir!" Carpenter saluted and as he left, he glances at Frazee and winked before heading out.

Dak interrupted, "Lt. Sharp. We have taken back the ridge. Enemy artillery and mortar fire has ceased."

2nd Plt. Btry. D-93rd AFA Bn. 6th Armd. Div. Ft. Leonard Wood, Mo.

Chapter Ten

The difference between a father and a friend takes many years to develop, especially for a son. Often some poignant event transcends the Parent-Child bonds to the friendship of an Adult to an Adult.

The transition for me took many small events over time. My father and I had many lunches, often with my Uncle Bob (aka Toby). At first I sought their advice on things then I began to listen. As I listened, I heard stories and tales that I had never heard before as they talked about their lives, desires, successes and failures.

Then one day we went and saw the movie Saving Private Ryan. Each of us had a different reaction to the different parts of the film. For me it was the bunker scene when Captain Miller looks out with the flash, whizz by the head, bang moment. For my uncle it was the detritus of war scenes. My father analyzed the different combat scenes tactically during the following lunch.

But, later I found out that one particular movie event really did get to him.

Right after the scene, where they secure a German machine gun nest near the remains of a radar unit on a hill and where one of Captain Miller's soldiers is shot, bleeds out and dies, my father had stood up and walked away out of sight for a few minutes.

It was a month or two later when my father dropped the bombshell that he had been in Korea in 1953 during the fighting. I had been aware of his extra curricular activities with the CIA while he was in the US Army and when working for the County of San Diego. I knew he had been wounded twice during his covert actions. But I was stunned because it was such an emotional description coming from a most rational individual.

What he mentioned to me was something that he had never mentioned even to his brother. My mother had never even hinted on it while she was alive. It floored me seeing him shaking from the memory.

The concussive wave that hit me then was not only that he was in Korea during the shooting, but he had been caught up in the last bloody

109

maelstrom of fighting with a ROK outfit just before the ceasefire. Simply put, his unit was testing a Top Secret radar and it had been cut off and surrounded by the enemy.

He briefly described the position, its emplacements and with his hands then pointed to where the forward observation post would have been. Then staring into the almost a half century past he mentioned he had held a fellow soldier who had bled out in his arms. Taking a deep breath he intoned that half of them died on that hill. With tears streaming down his face he solemnly reported to me that he had thought he would never ever get off that hill and then they had gone and made him an damned officer.

Over the next year or so he started revealing bits and pieces of the story to me and my uncle. But, it came with a price, vivid nightmares; because the floodgates once compressed and compartmentalized in one small area of his memory had been opened. He had never dealt with the horrible tragic event emotionally and it now nearly overwhelmed him.

Both my brother and I had been in the Army, but we had never dealt with anything like this. I had had to deal with my own trauma that I too had hid for years. In long talks with my Uncle, we agreed that my father's jagged pain had to be released and dealt with. And with my Uncle's guidance, we three talked about what happened in Korea.

At times I was cursed at, threatened by a few who saw me ripping open old wounds that obviously hurt my father. How dare I as a son do such a thing?

It was then that I as an old soldier once, realized that I was helping an even older former soldier deal with it, one soldier to another, one friend to another, one adult to another adult.

John R. Carpenter

In Sgt Park's quarters, Frazee was checking the coffee pot on Carpenter's stove. Airman Cushing and the medic Casper walked in. Frazee looked up expectantly and saw them smiling.

"They got the bleeding stopped and it looks like he will make it.

He will be going home. He told us to give you something," said Casper.

Cushing handed him North's flight bag. And Frazee reverently opened it and pulled out a cigar box. He lifted the cover and saw a full box of cigars. He handed Casper and Cushing two each. Cushing smiled and said, "There's more."

Frazee dug deeper and his eyes went wide. With respect he pulled out a dark colored bottle of Jack Daniels. "Before I left him, he told me to say thanks to the "officer" who allowed me to take him to the rear," proclaimed Casper.

The men chuckled at hearing Carpenter referred to as an officer. Coffee mugs were passed around partly filled with coffee by Frazee. Cushing did the honor of adding a dash of Jack Daniels to each cup. As the men sniffed and partook,

Carpenter and Whitehorse entered the room. Frazee thrust a cigar at each of them.

"A boy or a girl?" Whitehorse asked.

"Neither, my brave barbarian American Injun Warrior! My FAC Lieutenant's gift for being alive. Thanks to Carpenter, Casper and the mad driver here, Lt. North will get to see his wife and kids once more!"

Cushing smiled held up the bottle of Jack Daniels. Casper handed out mugs to Carpenter & Whitehorse.

Whitehorse responded. "Indeed! Better than a court martial!"

Whitehorse laughed at Carpenter who was momentarily hangdogish, until Casper and Frazee patted him on the back smiling.

"A Toast! To my Lieutenant! May he see his grandchildren's children!" said Frazee.

"Here, Here!" responded Cushing.

"Amen!" said Casper.

"Prost!" added Carpenter.

"Cheers! And may tomorrow be better than today!" finished Whitehorse.

Carpenter watched Corporal Bill Lipscome and Private Stanley Morgan supervise shell cleanup by the ROKs. Carpenter climbed up on the hood of the 40mm halftrack and looked around. He looked at the nearby shell holes created by the brief artillery barrage they had a few hours ago. "That was the first time I was under incoming artillery fire. And we are going to get more."

"First time for everything. And it was not that bad" said Lipscome.

"Do you think we should dig this in a little deeper? Maybe make another artillery shelter?" asked Carpenter.

"We are dug in according to regulations. This setup allows us to get out of Dodge when we are told to. If we dig in deeper, then we can't move em out. He pointed to the trench. "And we have the trench to hide in if the rounds come too close. Don't over think it."

Carpenter was still unsure. "Okay. Maybe you are right." He headed over to track 1 where Whitehorse was supervising the ROKs cleaning two of the M2 machine guns.

"Smart having the two remaining 50s ready to fire."

Whitehorse laughed, "SOP!"

"Hey, Chester. Can I talk to you?" Carpenter asked Whitehorse.

"Sure. What's up?"

The two men walked toward the rear and jumped the trench. Carpenter then looked back at Track 1. "We got our first real arty attack. And I think we should dig in better and build another shelter. What do you think?"

"Well, Dick. Since you have gone all informal and such. I agree. Matter of fact, I too was thinking the same thing."

"Really, Corporal Whitehorse?"

"Yes, indeed, Sgt Carpenter." He pointed with his chin. "See that stack of airfield landing mats?"

Both men started walking toward the broken open metal stack. Another one nearby still had the metal bands holding it together

"It looks like it got broken open by arty."

""Look closer."

Carpenter did and noticed the scratches and dings from the artillery fragments, but the steel mats were still usable.

"A bit dinged up. But the holes will allow shrapnel through. And, what am I missing?"

Whitehorse sighed loudly, "Dick. Dick. Have you ever heard of lamination?

"Nope."

"It is layered material bound together with other material to create a stronger bond."

"Ah! I get it. Steel mat, sand bags, cross mat, sandbag, then mat?"

"And sandbags on top of the top mat to avoid ricochets."

"H'mm. Who do we ask to get permission to use them?"

Whitehorse laughed again. "It is better to seek forgiveness. If they complain."

"Okay! Get a ROK crew on it. Dig in your track and build an arty shelter. And I will think of some good apology if we need it."

The ROKs then started filling sandbags and carrying the landing mats toward track 1. Later a ROK team was adding the metal mats in sandbagged layers over the trench to create an artillery shelter. They dug out nooks for sitting along the side of the trench.

In the early evening, an enemy artillery barrage walked into the position while the 1st squad 40mm was firing support to the ROK retreat. Immediately, the ROKs jumped off the track and ran for the safety of the trench.

"Come back here you cowards!" Lipscome roared.

The artillery bursts were coming closer and closer. Lipscome looked around and realized he was the only one on the track. "Damn it all to hell."

Lipscome was jumping off the track when one single artillery round dropped inside the track and exploded. The halftrack compressed under the explosion then rebounded upward making it appear to jump. The shock wave compressed the air solidly, creating a concussive force that slammed the airborne Lipscome off the sand bagged sidewall into a lifeless heap with blood oozing from every opening and his eyes bulging from the internal pressure. Ready ammo started cooking off as the track started to burn. The 40mm barrel, now bent, was pushed upwards at a near 90 degree angle.

Later that night, Carpenter was standing looking at the burnt out smoking remains of the 40 mm track. The first of the year's monsoon hit the area with wind, rain and lightning. The hissing sound of water on hot metal seeped into the night. Carpenter stared at the destroyed halftrack, a mixture of emotions on his face with each lightning flash that finally resolved into a determined look.

In the command bunker, Lt Sharp and his sergeants (Quincy, Carpenter and McCoy) were gathered together after the death of Corporal Lipscome.

"Track 2 is destroyed. I was able to salvage some gear. But the 40mm is kaput. Track 1 being dug in better, survived a near miss without any problem," says Carpenter.

"Track 1 was dug in better that Track 2?" asked the Lieutenant.

"Yes, Sir. I had them use some pierced steel planking in layers. I also had them build an artillery shelter in the trench."

"You stole airfield landing mats? Who the hell do you think you…" starts Quincy.

"Enough! Carpenter why didn't you have Track 2 dig in the same way?" again asked the Lieutenant.

Carpenter reluctantly responds, "I, uh, talked to Lipscome and he didn't think track 2 should be dug in better. Regulations are regulations."

"Great, save your own track and…" Quincy bemoans.

"Stuff it! Quincy, I…" starts McCoy.

"Knock it off Sgt. Quincy!" finishes the Lieutenant. Quincy was not happy and glared at McCoy and Carpenter.

"Our orders are to remain here. Like it or not. The fuel and oil created one hot burn once the ready ammo went up."

McCoy jumped in again, "Sir, We should also dismount the 50s. If we are digging in. It would be better to have our eggs spread out. We can build machine gun bunkers for the 50s with the 30 cal MGs in support."

"What about the remaining 40mm?"

"We would need an A frame with a block and tackle to lift it. And that's if we can undo the main bolts holding it in the frame."

"Quincy added, "On my next run to headquarters, I can check on the availability of a wrecker. They have an A frame and tools."

Sharps looked away from the men for a few seconds. He sighed, then turned back to them. "Drain the fuel and oil. Dismount the 50s and dig them in. We are going to get hit again sooner than later. And it is time to get serious about digging in. McCoy, you and Carpenter work with the ROKs. You guys are better at that. Dismissed."

As Carpenter and McCoy left, Quincy was held back by the officer. "Hold on a minute, Sgt Quincy. I want you to escort Lipscome's body to the rear, and look for a wrecker. I also need you to stop by the KMAG's office again. I have another letter to him and our ASA handler. We need to get out of here before it's too late. The radar can not fall into the hands of the enemy." Sharp handed Quincy the letter. Quincy turned to go. "And Sgt Quincy...See if you can find a can or two of fruit."

Quincy smiled at the request. "Yes, sir!"

A few hours later, a ROK crew began unloading a truck. New ROK replacements stood nervously with their weapons at sling arms. When the crew was finished, they loaded up the dead American, in a body bag, onto the bed of the truck. Quincy climbed in and the truck rumbled off. As the truck moved out, incoming artillery struck a few hundred yards away to the west. Then after a pause it struck closer. The young ROK replacements crouched and looked around, confused and fearful. Sgt Cho who had been supervising the work crew looked back at the replacements who were not moving.

"Those stupid green kids don't know shit!" Cho yelled out to the young Korean soldiers, "Get over here now, you stupid dogs! That artillery is coming our way!"

The replacements looked at each other, then looked where the Korean voice was calling to them. Suddenly they understood the danger they were in. They ran toward the trench and dove in. Carpenter was going down the back trench line past them when the shelling started up again. He smiled as he saw the ROK replacements running and diving into the trench ahead of him. As he reached Sgt Cho, he paused and saw the scared young faces of the ROK recruits. Most of them looked like teenagers. He moved over to Sgt Cho and smiled again. Cho reflectively smiled back. Artillery rounds exploded nearby. "Sgt Cho, it looks like we have new recruits."

"Yes. Very green. Like fruit. And with fear."

Carpenter looked at the young men. One looked so afraid he looked like he would run away. He pointed at that one ROK replacement and shouted, "You! What is your mother's name?"

"The American Sgt is asking what your mother's name is?" Cho repeated in Korean.

"Me? Who m,m, me?"

"Yes, you dumb shit. Your mother's name is?" Cho demanded.

"Her name is Flower," in Korean.

"Dumb shit says, Flower. Why ask?" Cho said.

"I don't want him running away. What favorite food does his mother make for him?"

What is your favorite food your mother makes for you?" Cho repeated.

"Eundaegu Jorim. A black cod and turnip soup. I like it cold."

"Cold? I like it hot and spicy!" Cho said with a smile. "He says Eundaegu Jorim. Fish soup. And he likes it cold!"

Carpenter looked at the ROK soldier then Cho. Cho was puzzled and not quite understanding, but faking it. The others are paying attention to the conversation, not the artillery. "Cold? That sounds yummy!" said Carpenter.

"He says it sounds delicious. But cold?" Cho translated. The men laugh nervously as Cho and Carpenter smile. Their confidence gives them confidence.

An artillery explosion nearby created a shriek of tortured metal. Carpenter paused and tilted his head upward and saw explosive smoke striking skyward. It was from the area of track 3.

Carpenter turned and pointed to the recruits. "Listen to your officers and more importantly to your NCOs! And you will stand a better chance at getting home to your mother's meals!" Carpenter paused as Cho translated. And before rushing off, he pat the shoulder of the nervous ROK shoulder and gave him a wink.

"He says to obey orders from officers. And most importantly from the Sergeants who will teach you how to survive. Then the odds are better to get you back to your mother's home cooking," said Cho in Korean.

Carpenter hurried down the trench. He now saw ammo cooking off and flames.

A few minutes prior, Corporal Lewis Brenner was under the 40mm mounted half track when the enemy artillery rounds started coming in again. He looked at the tough nut on the oil pan and decided to try once more. His wrench slipped and he smacked his wrist on the underside of the pan. He grabbed his wrist with his other hand and a rag partly covered up a nude tattoo on his forearm. "Damn it! I cut myself."

The incoming artillery started hitting a few hundred yards away. "It's starting again Brenner. Let's get to the trench," said Livingston.

Brenner listened again for a moment. "It is far away to the west. No sweat."

He smiled and grinned at the scared face of Livingston begging him to get out.

"No problem. I'm going to get that nut undone or else." Brenner then reapplied the wrench and with a grunt, spun the nut. He was victorious! He smiled and quickly got the oil pouring into his container. He started to crawl out on his back. Suddenly he looked puzzled as the half track compressed down upon him, then bounced upward and to the side from being hit from above by an artillery round. His nose began to bleed.

The front wheel came down and pinned his wrist, jerking his entire body in agony. His leg kicked the gas pan and gas flowed out over the ground around him. His other arm reflexively grabbed the oil containers, trying to pull himself away from the tire pinning his other arm. He heard someone screaming and realized it was him. He shut his mouth, biting his lip.

Brenner looked for help toward Livingston. He saw the man had rolled away and was pounding out sparks landing on the ground near him. Brenner's eyes went wide as he suddenly realized that there was a rain of sparks coming from above. He smelled smoke and tried desperately to get out from under the track. He couldn't get loose. He smelled the gas and oil and he started kicking dirt onto the ground near him trying to cover up the split gas and oil.

"Get me outta here! I'm stuck! Hurry up, grab my arm!"

Livingston quickly crawled over with his nose and ears bleeding from the concussion. He grabbed Brenner's arm and started to pull him out. There was a "whomp" and a flash of searing heat. Livingston stumbled backwards slapping out the flames on his uniform. Livingston looked back under the track and saw it was completely on fire. He spied Brenner jerking amidst the flames. He appeared to be screaming, "Shoot me! Shoot me! Please!"

Livingston heard nothing. He suddenly had his 45 pistol out and pointed at the figure of Brenner burning to death. Livingston felt the weapon fire in silence, and as one casing was still flying up in the air, it jerked again and again until the human figure was still and there were no more shells to fire.

Carpenter heard the shots and ran faster. When he turned into the 40mm half track bunker, he shielded his face from the heat. He peered around and saw an arm engulfed in flames and Brenner's tattoo was still visible in the flames. Carpenter looked away and saw Livingston still pulling over and over on the non-responsive trigger on his empty weapon. Livingston clothes were smoking and there was a terrible stench of burning hair and human flesh.

Carpenter grabbed Livingston and drug him away and down the trench line.

Livingston was still muttering. "I'm sorry. I'm sorry."

Chapter Eleven

Two Army trucks came roaring up the hill from the valley below. The enemy artillery started to fire, the explosions beginning to bracket the trucks. One truck seemed to surge forward but the other was flipped by an explosion and rolled off the side of the hill. Several men either jumped or were thrown from the vehicle.

The other barely slowed down as it came by the rear of the bunker complex. SFC Quincy jumped off with a few boxes. He hit the ground and then rolled, then dashed for the safety of the trench.

The army truck dashed down the airstrip as if it was ready to take off, heading for the western bunker complex. The incoming artillery rounds tried to follow it. Quincy scrambled out of the trench and retrieved the boxes with the help from a few ROK soldiers. A short time later in the command bunker, Lt Sharp and his sergeants (Quincy, Carpenter and McCoy) were present.

"What a cluster fuck it was. Nobody, and I mean nobody knows or is willing to do anything about us. I got some sympathy. I got some pears," Quincy said sadly. He handed the cans to Sharp. "And some coffee." He handed a can each to McCoy and Carpenter.

"For the men, some 10 in 1 rations. Not much, but it beats rice and spam."

Lieutenant Sharp spoke up, "While you were gone, we lost Corporal Brenner. He got pinned under Track 3. Livingston tried to pull him out, but…"

"But what sir?"

"He caught fire and Livingston couldn't pull him out," McCoy added.

"Oh, my God! He burned to death?"

"A little, then Livingston uh, well…" Lt Sharp stammered.

"Livingston put an end to it," interrupted Carpenter.

"Shit. A hell of a thing ... How is Livingston?"

"Casper says he is in shock. He has some burns, but not enough to justify that he be sent to the rear. Later tonight or tomorrow, the ROKs will take Brenner's body and a few other ROK KIAs to the rear. They go after the wounded," said Sharp."

McCoy changed the subject, "Sir, Here is a sketch we made for disbursement of the 50s and 30s. Giving us an interlaced defensive arc of over 175 degrees. To the rear we have posted a pair of 30 cal MGs."

"How is the radar going to help?"

"Carpenter figured it out. Carpenter, you describe it," said McCoy.

"We have 2 radars for each side. One covers each flank and the other two sweep the front. We place one radar with the center of three 50 calibers for the front and one with the 50 on the flank. If needed the forward facing radars can command all six 50s facing forward through our communications set up. The flank radars control their 50 and one more if needed."

McCoy again piped in, "To make a long story shorter, Carpenter has used 40mm casings as aiming markers. For the 50s adjacent to the radar, he offset the aiming stakes to match pointers of the central 50. Crude, but it should work."

"So as long as we can maintain internal communications, we can use the radars with the 50s. So we are back in business with the radar operators able to give direction, range and elevation to allow three 50s per side to converge on target. Very good. How is the digging in going?" said Sharp.

First squad has a jump on 2nd squad. But the basics should be finished by tomorrow," said McCoy.

"Just get it done. My guess is that we don't have much time. See to your squads. Good job you two," Sharp exclaimed.

"Thank you, sir," said McCoy.

"Yes. Thank you, sir," added Carpenter.

Carpenter and McCoy left. Sharp turned to SFC Quincy. Quincy stepped to the back side of the command bunker with Sharp. "That KMAG officer is a prick. Pardon my language, sir. But he told me if we come off this hill without orders, he will see that each one of us are tried for desertion. He also added, that is, if we are not shot by the ROKs first."

"Unbelievable. Because of secrecy we were put here. Because of secrecy we can't leave. We are lost in limbo on this damned hill."

"Yes, sir. We are in one hell of a pickle. Some of the men have begun calling the detachment the lost bastards. The only positive is the KMAG officer agreed to send your letter to 8th Army Command MI describing our mission and that our contact is M.I.A. Maybe the Military Intelligence boys will get us out of here. I got a bad feeling about all of this."

Carpenter and the other men finished rebuilding the machine gun nests and placed the 40mm casings as aiming stakes. The men continued to cover the trenches and placed airfield metal planking in the bottom to replace older wooden duck boards where water had gathered from the off again, on again rain. The communications network was tested.

A squad tent was placed over the ramp to the downstairs bunker. Men then lined it with sandbags and sandwiched pierced steel planking with more sandbags to prevent water from running downstairs. One of the men asked another a question, "Did Col Kim authorize the use of this?"

"No. The American officer did I think. Lt. Dak and the NCOs are not complaining. So it must be alright."

"So if Kim gets mad, Lt Dak will blame the Americans!"

"Blame the Americans!" Yes, That is good!"

McCoy and Carpenter were inspecting the bunker complex; when they neared 2nd Squad's last surviving half track (Track 4), Riney was using the dismounted generator to charge the vehicle's batteries. A few other Americans were nearby. "What's going on?" asked McCoy.

"We are getting this track ready to move out. When orders allow us to bug out, we will be ready,"

McCoy glanced around the now mostly enclosed gloomy bunker

and snapped on his flashlight. He spied several 5 gallon jerry cans of fuel nearby. Riney started to say something, but McCoy turned on toward Riney. "We have orders to stay put. When and IF it is time to go, I will tell you. Get that gas out of here!"

"That is bullshit. You and I both know the ROKs will bug out when the pressure gets bad. They will hang us out to dry. Just watch and see!"

"Move that gas out of here. It endangers everyone here. Do it now. Move it!" A few men slowly picked up the gas cans and started moving them out. Riney looked like he was ready to fight. "If I hear you talking about bugging out again, I will kick your ass all the way back to the stockade."

"At least in the stockade we eat American food and we will all be alive."

"Riney, I always thought you were a fighter. Now, all I see is a quitter."

Riney checked himself then kicked an empty pail against the track. It made a racket. Without another word Riney left with both fists clenched. McCoy turned to Carpenter with a worried expression, "We need to keep an eye on them."

Carpenter walked over the hood of the track and removed the safety hatches for the hood. He lifted the hood and McCoy held it up. Carpenter removed the distributor wire and stepped back. McCoy set down the hood and latched one side while Carpenter did the other side. "Here." He handed the wire over. "For better or worse, we stand here together."

Col. Tang, Regimental Commander & political commissar of the 203rd Division, used an old ROK slit trench (open toilet) near a bunker his men were renovating. After he had finished his business, he and his body guards moved toward an observation position. There he used captured American binoculars to view the ROKs digging in below them in the valley.

A ROK sniper with a bolt action M1903 rifle, with a highly

prized 30x scope, fired a shot and noticed his round falling very short of the CHINCOM officer using his morning toilet.

"Too far away. He did not even notice it. That was a wasted shot! But, someone needs to bag that pompous bag of shit!"

"Maybe you just need a bigger gun?" came the snide remark from the other ROK soldier lying next to him in the sniper nest.

"You may be right."

"That you need a bigger set of balls with your gun?"

"No. A more powerful gun. And I heard a story of an American who might help." The sniper looked back toward Hill 433 smiling.

Carpenter watched a ROK soldier with a very large scope approach ROK Sgt Cho. He got up and started to amble over.

"Sergeant, I am looking for an American shooter who can help me bag a high level Chinese officer," said the Korean sniper to Cho.

"He must be impressed by your fine rifle. Here he comes now." Both ROKs turned to Carpenter walking up. The sniper looked at Carpenter critically. Carpenter was patient, he looked at the rifle and scope while they talked about him

"Him? He is the shooter?"

"Yes. I lost good money when he split a bullet 3 times over an ax blade breaking condoms used as balloons."

"He must have been close."

"15 steps. I measured it."

"Sergeant, would you ask him?"

"This man a sniper. He no hit big chink officer. Too far away," Cho said to Carpenter. Carpenter gestured to the rifle.

"May I see the rifle?"

The ROK sniper cleared his bolt action rifle and handed it to the American. Carpenter began to smile as he looked over the weapon. It was clean and well cared for. He looked down the barrel, checked the action and scope. "If you can't hit him with this, then you need a bigger weapon."

"He says you need bigger gun," said Cho.

"I get no respect... Yes, I agree, but how can he help?"

Cho turned to Carpenter, "But, with what?"

"M2. That should do."

"A M2 machine gun? Not accurate," said Cho.

"M2? Machine gun? 50 caliber that would be big enough! But how?" the sniper asked.

"He ask. How?"

Carpenter tapped the 30x scope. The men walked away with Carpenter gesturing. Carpenter then drilled and tapped the scope to the top of an M2 machine gun in the bunker below. He added a post sight at the base of the double trigger, to allow firing the M2 as a single shot weapon above ground. The ROK sniper was enlisted as Carpenter's spotter. The added post sight allowed the same sight picture every time. Adjustments were made. Then the ROK sniper was grinning with Cho & Carpenter.

Before dawn the next morning, Carpenter was in front of Hill 433 in the valley northward. He was setting up the scope mounted M2 MG (machine gun) to fire single rounds. All was quiet and peaceful. The ROK sniper tapped Carpenter's shoulder and pointed to his wrist watch. Carpenter nodded and turned the weapon towards the distant slit trench. He was very focused on his goal.

In the big scope he saw the Chinese officer, Lt. Col. Tang, come out for his morning toilet, with a yawn. His bodyguards moved to guard him. The cross hairs moved about 5 feet over his head. Tang undid his trousers and assumed the squat position to void his bowels. For a moment, one body guard blocked the shot. The cross hairs gave a slight wobble as it was adjusted, then it was steady. The valley saw a flash of light then heard a loud bang as the M2 fired. The scope steadied again and the snipers saw a brief contrail, then a burst of bloody haze as Tang's body was blown into the old ROK slit trench (toilet).

For the next three days we got some artillery and sniper fire but no large scale attacks. One thing that did happen however was an experiment I tried. There was a Chinese Officer who came out on the ridge line across the valley to our North every morning and took a crap. After the second morning of watching him, I fixed a Sniper Scope to a 50 Cal. Machine Gun, and zeroed it in on some rocks at the same range and several hundred yards to the East of where he came out. Then I waited, and sure enough he made the mistake of coming out to relieve himself on the third day, at the same time-I was already sighted in on the position, and there was no detectable wind. I took aim, and fired. He simply disappeared, thrown over an embankment when the round hit him. I bragged about that shot all day long.

Richard Carpenter, Memoirs

Chapter Twelve

July 13, 1953

The Chinese attacked the ROKs with artillery, mortars and human wave attacks during the night. Some positions were overrun and the defenders were all killed. The pressure mounted and the ROKs fell back orderly. Then one or two began to run. Then others ran. Kim was seen yelling orders then reluctantly falling back also. Sgt Park watched his back.

U.S. Army 555th Artillery Battalion at Ich'on-dong
Near Highway 117A

US Army artillery men were firing their 105mm weapons when a ROK soldier ran up yelling. The ROK was pointing at a nearby hillside. In the dark they could not see anything. Suddenly a parachute flare erupted over head casting its light on a hoard of Chinese coming down the hill toward the American position. With much confusion and yelling the Americans hurriedly man handled and turned two of the large cannons to fire point blank at the enemy. Americans started falling.

One weapon fired a canister round (like a shotgun round) and many enemy fell. The other weapon pointed at the enemy and an American grabbed the lanyard to fire, but was struck down by enemy fire. Another American grabbed the lanyard to fire and he too was struck down. Then the Americans were overrun. Many tried to surrender. While many were still shot, many were taken prisoner.

HILL 433

Kim and a handful of men followed him into the FSB on Hill
433. Sgt Park directed the men toward defensive positions. Kim
entered the command bunker and saw Lt. Sharp. Sharp and Quincy
were listening to the cries of an American Artillery unit. They heard
the following, "They are everywhere! Mother of God! This is the Triple
Nickel, we are being overrun! We are being over…"

The sound of automatic weapon fire cut the scream off. Then
there was static.

"It sounds like one cluster fuck out there. What is going on,
Col. Kim?" asked Lt. Sharp.

"My Battalion is falling back under pressure. We will rally here.
Get all of your men ready to fight!"

"Yes, Sir!"

Sharp reacted to the command and threw a quick salute as he
left the bunker. Quincy, looking troubled, followed.

The Korean radio operator then spoke, "Col. Kim. Radio
messages are confusing. The enemy is jamming many frequencies.
On the emergency channel a fall back to the Kumsong line has been
ordered." Kim looked at the map.

"That does not make sense, unless the enemy has broke through
in force along the flanks. And … that means we must hold as long as
we can. This will give them more time to ready a counter-attack." Kim
then looked directly at the radioman. The man fidgeted under the stare.
"Did you tell anyone about that order?"

"No, sir. Lt. Dak left just before it came in. And my English is
poor, so I did not tell the Americans."

Tell no one about the fall back order. Do you understand? Not
even Lt Dak."

"Yes, Sir. I understand and I will obey."

Kim then looked at the soldier. He laid his hand on the man's
shoulder. If I have a chance, I will explain later. But, know this, it is

very important what we do."

The man nodded respectfully. Kim gave him a slight bow then started to leave the bunker as Lt Dak came in.

"Sir! I have all NCOs out gathering as many soldiers with weapons as possible into our perimeter. We even have some of the 6th Division coming in."

"Very good, Lt. Have the cook warm up what he can. Set up relays to bring food, water and coffee, if possible, to the men. I need them ready to fight!"

Kim left the bunker and headed out into the open. He could see men coming to him and being directed by Sgt Park. Kim ran down the back trench line to the west side of the complex. When he got there he was amazed. He saw the American Corporal Whitehorse smiling with a ROK soldier as they waved ROKs into the trench lines. He passed them and then saw Carpenter standing next to Sgt Cho. He heard Cho yelling at a group of armed men trying to bypass them.

"All those with weapons! Take positions to my right!" The men ignored Cho and kept walking."

"Chickens! Dumb shits!" yelled Carpenter. Then he said in broken Korean,

"Dumb shits!"

The ROK soldiers stopped to look at the American upon hearing English. Then they laughed at his attempted insult. They walked over and touched the American as they passed by, headed into the trenches.

Early the next morning, Carpenter, Whitehorse and ROK Cho watched to the west. The sounds of combat and flashes of grenades with slightly delayed thumps came from the western end of Hill 433. The other ROK bunker complex on the west side was overrun. After a small explosion and fireball, the battle noise died off. Single shots rang out, then silence.

"Single shots? What is going on?" asked Whitehorse.

"They call it mercy. Those wounded too bad. Ours and theirs," responded Cho.

"Remind me not to get wounded. How long before they come for us?" asked Carpenter.

"Soon. Very soon," said Cho.

"Probably at oh dark thirty," added Whitehorse.

"Then it will be at oh shit thirty," said Carpenter.

Carpenter headed down the trench, passing a ROK 30 caliber position. He did not recognize the men, but they smiled at him. Two reach out and pat his arm. Carpenter looked embarrassed, but nodded politely and moved on. He entered the 50 caliber position where Ransom was stationed, flanking the left center radar. Several ROKs were there. "Thumbs, (nickname from Ransom's clumsiness when excited) you know what to do. Remember slow and steady."

Carpenter smiled. Ransom looked scared but nodded his head. The ROKs nodded also. Carpenter continued his check by going into the radar bunker.

"Smittie, how's it hangin?" he asked.

"One a little lower than the other."

Carpenter smiled, left the position and saw Tucker showing an ROK soldier how to rotate the radar up just above the sandbags. "Hey, Tucker! Good idea!"

Tucker looked down and smiled as Carpenter went by. Carpenter then entered the other 50 caliber position flanking their radar unit

"Morgan. All is well here?"

"Yes, sir!"

"I'm not an officer. Don't call me sir."

"Sorry Sarge. All is okay."

Carpenter then headed toward the rear trench line. He joined a group of ROKs heading toward their assigned fighting positions. Carpenter went by a few 30 caliber positions, asking "All ok?"

"OK, Joe!"

Carpenter saw ROK Sgt Cho. The man was eating some dried fish and smiling. Carpenter had to smile back. "Keep your eyes open and watch our rear!"

"Roger dodger," Cho answered.

Carpenter laughed at Cho's choice of American slang. He patted the man on the shoulder. As Carpenter left, a few ROK stood and reach out toward him. Carpenter stood still, uncomfortable as they briefly touched him. They smiled, then gave a slight bow, as if slightly embarrassed. Carpenter returned the slight bow and moved on.

He reached the far left flank 50 caliber gun position and saw Whitehorse already checking the weapon. He stopped for a moment.

"What is it with all this touching?"

"Huh?" Whitehorse answered.

"Why do they keep touching me?"

Whitehorse turned and looked at Carpenter bemused. He reached out and touched Carpenter lightly on the arm. "You mean like that?"

"Yeah! Just like that!"

"They think you are lucky. And when they touch you, they think it will get them some luck." Carpenter shook his head and went into the radar bunker.

Jose Rodriguez was watching and listening to the phone. He turned toward Carpenter and remarked, "They're lining up toward the north. It is amazing seeing this. Scary but amazing!"

"Modern wonder. What will they think of next?"

Carpenter turned toward ROK Private Chin Ho Sun. The little guy was fast asleep. Carpenter used his foot to nudge Sun's foot. Instantly Sun was awake with eyes on Carpenter. "Good morning. Pass the alert." as he gestured toward the north.

"The enemy is coming from the north."

"Alert. North. Yes!"

The young ROK jumped up and ran out the door to warn everyone.

"Smart little guy. I hope he makes it," said Rodriguez.

"Me too! Me too!" replied Carpenter.

Carpenter went out and looked up toward Smith rotating the radar manually.

He had a ROK soldier watching him. "Smith! You've been talkin to Tucker?"

"Of course!"

"Good job, GI!"

Carpenter headed out and went past the flanking 30 caliber position. He heard the bugles start their off key sounds. He then returned toward the left flank radar position.

"Carpenter! Here they come! Less than 400 meters!"

Carpenter stood up and climbed the sand bags next to Sam Smith. Using his binoculars, he saw nothing out there. He slid down and readied his flare gun. Carpenter heard Rodriguez reporting on the power phone the enemy is coming.

"Roger. Large mass of infantry to the north. Now at 375 meters. Out."

Rodriguez looked at Carpenter worriedly. He checked his screen again, still holding the phone. "350 meters."

"Smith! You be careful up there!" Carpenter yells. Smith looked down from his radar rotation position and gave Carpenter a thumbs up. Carpenter aimed his flare gun to the northern sky and fired. He heard the enemy whistles and a different bugle call. "Okay, Jose. Tell the M.G.s to open fire as planned."

"All 50 cal positions open fire as planned!"

A 50 caliber machine gun began to fire. Long tracers reached out toward the north. Some hit the ground and flew away on a different trajectory. The other 50s opened up. The flare burst and shed its light. The Chinese human wave was seen approaching at a jog up the hill. In the flare light and tracer light the men saw the enemy falling, the gap close, but they were still coming toward them. Carpenter readied another flare.

"Carpenter! Where is our artillery?"

"The net is down. I guess Frazee crapped out."

"Fuck him and his Air Force!"

Carpenter looked at Smith and shook his head.

"Carpenter! 275 meters!" yelled Rodriguez. "250 meters!"

Carpenter then fired another flare up in the air toward the enemy. In the sudden light, the enemy was seen and the 30 caliber machine guns started up. The 50s shut down and the crews madly began to replace the hot barrels. From the right another flare went skyward replacing the one Carpenter fired.

"200 meters!"

As the 50 caliber machine guns came back online and started firing again, Rodriguez, focused on his radar, started to look puzzled. Then he looked toward Carpenter and shouted, "175! ... gaps ... 175 and ... 200! They are retreating!"

Carpenter scooted up and looked toward the enemy and confirmed with his binos that the enemy was retreating. He then slid down and yelled, "Cease Fire! Cease Fire!"

Rodriguez used the power phone to relay the order. The machine guns stopped firing nearby. However, the rear 30 calibers were firing. And Carpenter heard another sound, an automatic weapon sounding its "burrrping" fire. Defensive rifle fire and grenade explosions were heard from the rear.

"Shit! Rodriguez! Sound the alarm! We are about to be over run from the rear!"

Carpenter looked out over the trench to the south; he saw grenade flashes, then Sgt Cho's 30 caliber tracers stopped firing.

"Crap-in-nola! Everyone but radar people listen up!" Carpenter turned toward the ROK soldiers and the few Americans. He grabbed a BAR (Browning Automatic Rifle) and a bandoleer of 20 round magazines. As he slapped a magazine in and worked the bolt to seat a round, he flicked off the safety and yell, "On me!" Then in Korean, "Follow me!"

Carpenter dashed off down toward the back trench followed by the ROKS. As Carpenter got to the rear he saw three Chinese soldiers within the perimeter firing downward with their Russian burp guns. He lifted his BAR to his shoulder and took them down. He fired a few more rounds then ducked to reload. After smacking the magazine into place, he worked the bolt and was off running. The ROKS ran, fired, ran and fired behind him.

Carpenter saw several dead ROKs and headed to the 30 caliber position which was silent. He put in a fresh magazine. A grenade came flying at him from the trench ahead and he bat it away. He kept going, scrambling forward toward the enemy in the trench. He heard the grenade explode off to the side and when a Chinese face looked to see the damage, he fired a burst, but missed his target who ducked away.

Carpenter emptied the BAR magazine, as he charged forward; into the sandbags protecting the enemy. The heavy 30-06 rounds chewed up the bags. He then butt stroked around the corner and smacked the figure there in the face. He struck again and blood spatter got on his face. Several ROKs charged past him down the trench.

Carpenter reloaded, seeing more enemy above ground. He sprayed automatic fire from the BAR, then tossed a grenade as he ducked to reload. The grenade exploded and he was off running again. He passed wounded ROKs and those guarding them. One yelled at him as he goes went by, "OK, Joe!"

Firing here and there at enemy soldiers, Carpenter saw Cho's 30 caliber machine gun bunker just ahead. Two of the quilted, jacketed enemy were outside. From the hip Carpenter fired, mowing them down. Suddenly the enemy was in retreat. The attack was broken. Several ROKs ran by Carpenter to get their last shots in. It started to rain with slow big drops at first. Carpenter reached the bunker and yelled inside, "Smiley (his nickname from Cho's huge smile)! Cho! Can you hear me?" Carpenter waited for a response and kept the entrance covered. Then a voice was heard.

"Carpenter? What? Why here?"

Cho stuck his head outside. Carpenter pointed with his BAR at the dead enemy soldiers. Cho looked then starts smiling brightly.

"You nearly got infiltrated, why are you smiling?"

"Why? You here! You are Number 1 GI!"

Carpenter lowered his BAR. Then he started to smile too. The rain was warm and began to soak them. Both turned toward sudden movement up the trench line. Carpenter's BAR came up. Suddenly,

Whitehorse was there carrying an M1 rifle.   "We beat them off!" he yelled.

"Damned right! We beat them off."

Chapter Thirteen

Carpenter looked out at the shifting fog and rain. He once again checked his positions on the western side of the defensive ring. He paused to check on some ROKs. They were digging in more and used dead Chinese as part of the new positions. Here and there he saw body parts being covered. He heard Korean Folk music being played in the distance.

"Ammo, Grenades, water okay?" he asked.

"OK, Joe!"

Carpenter pointed at himself, "Me, Car-pen-ter. Not Joe."

"OK, Joe!"

Sgt Cho came up and looked at the "OK, Joe" soldier disgustedly. Then he smiled. "He speak two words American. That all. But he good fighter."

"OK, Joe!" the younger Korean soldier said again.

Carpenter grinned and gave the ROK a thumbs up. Then he turned back to Cho. "What is that music?"

"Ah. Korean Folk music. Good sign. Chinks no attack, music play."

"Good. How are your men?"

"We dig. Much ammo, grenades and…" He held his hand open to the rain.

"Much much water!"

Carpenter smiled and patted Cho on the jacket and moved on. The fog shifted here and there. Sometimes visibility was only a few feet, then a hundred yards or more. Carpenter walked into the bunker where Rodriguez and Smith were watching the radar.

"That ROK is too jerky. Go talk to him again," Rodriguez requested.

"Okay. I will try again," said Sam Smith.

Smith left and Carpenter heard him climb the sand bags. He heard him say something in Korean, then a smack noise and a single distant shot was heard, then a tumbling sound. Carpenter and Rodriguez jumped up and ran outside. Sam Smith was crumpled at the bottom of the trench. Carpenter ran over and turned Smith over. His head flopped. Smith's eyes were open and there was a gaping hole in the back of his neck. He was dead. Carpenter looked up and saw a scared Korean hunkered down just below the top of the sandbags. The Korean pointed at the radar and there was a hole in it.

"Shit!" says Carpenter.

"Oh, Dear Lord. A sniper got the radar and Sam at the same time," added Rodriguez. Carpenter turned away and he slowly closed the dead man's eyes. For a few moments, Carpenter began to cry, then he heard Ransom calling out to him. He wiped the tears away.

"It is okay. I'll take care of him. We'll take him below and get him out of the rain."

Carpenter wiped his eyes and nodded his head. He watched as Ransom and a ROK soldier carried Smith away. Slowly Carpenter got up and looked up toward the shocked ROK soldier who looked stricken, still holding the broken radar. Rodriguez came out of the radar bunker.

"Now we are really FUBAR'd. The radar is dead, too."

"Report it KIA, and try to fix it. Have Tucker bring his radar over here," Carpenter ordered. As he turned away, he shook his head, "Shit, shit shit!"

Carpenter motioned for the ROK to bring the radar down. As he did so he pointed at the hole and asked, "Shit?"

Carpenter nods his head, "Shit. Double and fuckin triple shit!"

1930 hours

A ROK on lookout yelled, "Movement! Flag!"

Carpenter clambered up and saw a waving white flag. Whitehorse joined him.

"A white flag from the Chinese? Do they want the to surrender?" Carpenter asked.

"Doubtful. The loud speakers have had pleads for the ROKs to surrender."

Carpenter called down, "Rodriguez! Call it in. It looks like they want to parley."

A short time later, as twilight approached, Lt. Col Kim and Senior Sgt Park arrived. They both climbed the sand bags. Carpenter handed Kim his binoculars. Kim scanned the two Chinese. One was obviously an officer. "A high ranking officer. That means they are serious. I will go meet him," Kim stated.

"What if, it is a trap?" Carpenter responded.

"Such traps work both ways. I have already ordered a sniper to be ready if I fall." Kim jumped down and motioned to Sgt Park. "I will need your eyes on this, Sgt Park."

Park nodded, then lead Kim down a patch to the wire carrying a pole with a white T-Shirt on it. Two ROKs with gloves moved the wire to make a path. Park waved the T-shirt flag. Following a path marked with bits of tape, the two zig-zagged through the mines. They went out to the Chinese, who had begun to walk forward to parley. The men walked past bodies, parts of bodies and torn up ground. After another hundred yards or so Kim could make out the CHINCOM officer. He recognized the Chinese rank of senior colonel. Kim trying to be proper, stopped a few feet away, then saluted him. Colonel Zhou returned the salute and introduced himself. The conversation was in Korean.

"I am Senior Colonel Zhou, 203rd Divisional adjutant of the 68th Chinese People's Volunteers. Please pardon my weak Korean language skills."

"I am the commander of the bunker. Your pronunciation is adequate. What do you want?"

"No rush, Lt. Col Kim. It is you, is it not?" He smiled. "Would you like an American cigarette?"

Kim held his poker face as he took one of the cigarettes. Sgt Park stepped forward and used his American Zippo lighter to light Kim's, then Zhou's cigarettes. Then he stepped back to his spot. The two officers appraised each other. Zhou inhaled deeply enjoying his smoke. "Fresh American cigarettes. The spoils of victory. Shall I cite your birth and father's name to convince you that your prowess precedes you, Lt. Col. Kim?"

"No need." Kim makes a show of enjoying the cigarette while continuing to study Zhou.

"You know by now that you are surrounded. Most of your vaunted Tiger Division has been destroyed. What is left...What does your American puppet masters call it? Ah, yes. "Bugged out.""

"But, here we still stand," Kim replied icily.

"And I congratulate you for that. Your White Tigers did fight well. I lost many many men to you. Of the 21 strong points in your sector, you are the last. And because of this, I will be generous. Surrender now and you and your men will live. And within a few weeks or months you will be honorably exchanged when the fighting will cease."

"And if I don't surrender my command?"

"Oh. I am so disappointed that you think to be unreasonable. I will even allow your Sgt Park here, to continue to be your aide, and…" He smiled a sly smile.

"After your surrender, even reunite you with your family." The Chinese officer let that statement sink in with silence.

Kim's demeanor broke a bit. He was caught off guard. "My family? Where?"

Zhou nodded, realizing he was getting the upper hand. "Your wife is so much better after being injured by American bandit aircraft. I have personally made sure she is being treated well. Along with your young son and beautiful daughter, of course."

Kim tensed and allowed a little anger to come out. "Where?" he asked firmly.

"Why in Kumsong, of course. Your old home is being used by the good People of North Korea, so she is a guest of the Chinese

People. I will be happy to see you reunited with your family...after your surrender."

"I will have to think about this. Maybe you are lying."

Zhou reached into his front pocket and pulled out a gold wedding ring. He handed it to Kim. "She was reluctant to give it up. She said she misses you."

Kim held the ring up to see the inscription inside. In Korean it had his wife's and his family names. He took in a deep breath. His fist closed over the ring. It was his wife's ring.

Zhou believed he had him. He gave a tight smile. "She will be very very disappointed if I tell her you chose not to join her...*very* disappointed."

"I need to talk to my men."

"You have until first light. Until then enjoy the music. " The Chinese officer sighed a long sigh. "Then so be the fortunes of war. Keep the ring as a reminder, Lt. Col Kim." Kim slowly saluted the man. And Zhou smiling returned the salute, then turned away, walking back to his lines.

Kim looked at Sgt Park, for a moment he looked stricken. Then with a deep breath he headed back to his command. As they walked carefully back. Sgt Park turns to his commander, "He is lying about your family."

"I know. He did not provide the code word from her. And the ring has cut marks on it."

After Kim and Park reentered their defensive position, they heard a distant shot of a pistol. Then another. Carpenter shook his head sadly and lowered his binoculars. Lt Sharp was now present. "That Chinese officer just shot two of his own wounded out there," Carpenter commented.

"Mercy killings. I have seen them do that before," Kim replied.

"Colonel Kim, What did they have to say? Do they know about us, uh, I mean we Americans?" Sharp asked.

"The Comrade Senior Colonel, similar in rank to an American Brigadier General, says we have until dawn to surrender. And he made no hint of knowing that any Americans are present here. Otherwise, he would have made sure to mention it."

"Sir, And if we don't surrender?"

"Then they will try to annihilate us. The good news is that they do not know your secret. That is an advantage they will not expect."

"That does not sound like very good news."

That means they will underestimate our readiness, commitment, and resolve. That will work to our favor." Kim, Sharp, and Park left the bunker. Overhead Chinese artillery rounds were seen shredding the clouds above. The Korean music continued playing for the rest of the day with occasionally pleading calls to the ROKs to surrender.

The music continued to play. The ROKs were not working; they were listening. Some were reflective, a few cried softly. Those working on digging in occasionally stopped before others reminded them to keep working. Sgt Park observed this. And he reported such to a command meeting. Sharp and his three Sgts were present.

"Sgt Park. Your report please," commanded Kim.

"Too many are listening to the music. Not too many of our unit, but others, less well trained. The music is bad," said Park.

"I agree. If we were not surrounded and did not have so many untrained men, then it would be a different story."

"If anyone deserts and are captured, they will reveal that we have radar here. So, how can we help, Colonel?" Sharp added.

"We will send out patrols and locate the loud speakers. Then we will stop them," said Kim.

Several patrols were sent out. All confirmed that they were surrounded and one found where the music was coming from. The loudspeakers were set down a nearby cliff and guarded. Carpenter and Park were sent out. Their men rolled a barrel of gasoline. Carpenter mounted an explosive charge and rolled out the wire. When he set off the charge the barrel was tossed up in the air and started down the cliff. A second charge blew the barrel apart and ignited the gas before it hit the ground. The loudspeaker station was drenched with flaming gas. A loud feedback and devilish screech was heard before the large loud speakers died.

16 July 1953
0545 to 0845 hours

Another assault was fought off with artillery assistance. In the command bunker a call came in on the power phone. Lt. Dak was responding. Lt. Sharp was at the observation slit.

"Say again? Roger. Enemy lining up at 600 meters to our north per radar. No visual due to dense fog. Out," Dak hung up the phone and turned to Frazee. Frazee was already working the radio.

"This is Bluefin 4, Bluefin 4. Any station. Repeat any station. We need fire support, over."

Static was heard on the radio. Dak looked at two ROK runners standing by,

"Sound the alert! Go!" The two runners were out the door quickly. Dak turned to the ROK communications NCO, "Alert all bunkers. Enemy lining up for attack."

The man nodded and quickly cranked the power phone and started notifications. Just then he heard the static give way to a voice. It was an 8 inch battery replying over the radio.

"Bluefin 4, this is Rancher. Over."

"Rancher, this is Bluefin. Fire mission! Over."

"Bluefin 4, identify F.O. Over."

"My six was North. My 4 is fish. Over."

"We were told you were dead. Over."

"My North is in Japan. I am fishin for gooks! Over."

"Roger Bluefin 4. You got a target for us? Over."

NOTE: When giving numbers over radio 44 was 4 (four) 4

145

(four) not forty four. Between number sets was a slight pause marked with a comma. This was arty fire without preset target locations.

"Thank you Rancher. Target in the open. Map Grid 71,44. Repeat 44,71. Over."

" Standby." Many seconds went by. Then after a crackle of static, "Bluefin. Ready for fire mission. Over."

Frazee looked up and Dak was ready with numbers. He handed them over.

"7120,4425 Repeat 7120,4425. No smoke. Concentration only. I will adjust. Over."

"7120,4425 No smoke. Concentration only. Hold 1... Outgoing! Over."

Frazee looked up toward the ceiling. Then a rumbling noise was heard as two 8 inch shells rumbled over toward the target.

"Those are 8 inch shells!" said Dak. They heard the two explosions in the distance. There was a jangle of the power phone. Dak started to pick it up but Frazee beat him.

"Command ... Oh, Carpenter. Yes? One hundred fifty long and one hundred right? Standby!"

Frazee with a jump was back at the radio. "150 long, 100 right. Concentration only. Over."

"150 long, 100 right. Standby...Outgoing."

"Outgoing!" The rumbling noise was heard again then distant explosions

"On target! Fan-damn-tastic! Standby," said Carpenter over the phone.

Frazee jumps back to the radio "On target! Fire for effect! Over."

"Fire for effect! Standby ... Outgoing! Over." The rumbling noise was heard again, this time it was like a freight train as the large 8 inch rounds were fired from all the guns in the battery. Distant explosions were heard. They started to hear the 50 caliber machine guns firing.

Frazee was back on the phone "Outgoing! ... Roger! Those are air bursts. ... What? Drop 50. Roger. Standby."

Frazee went back to the radio. "Rancher. They are still coming on. Down 50 then down 50 on next barrage! FFE! Over."

"Down 50 then down 50. FFE. Standby ... Outgoing! Over."

Frazee goes a bit slower back to the phone, "Outgoing!"

The freight train came over again with a pause and distant explosions. Then came the freight train again with closer explosions. The 50 caliber machine gun fire stopped one by one. Frazee looked around then listened on the phone.

"Enemy in retreat. Roger! Out."

Frazee tossed the phone to the ROK operator and called back over the radio, "Rancher! Enemy retreating! Thank you! Bluefin 4 out."

"Bluefin! Call us anytime. Rancher out!"

Frazee got up and stretched. He walked over to Lt. Sharp by the observation slit and looked out. There was a thick fog.

"I could hear, but I didn't see a single flash. Do you know what you just did Sgt Frazee?" said Lt. Sharp.

Frazee turned to Sharp slightly puzzled.

"You just conducted the first combat use of radar directed barrage fire in the history of the United States Army. Well done!"

"Well, ya know all that jumpin back and forth an such, just about tuckered me out. That is for one of them darned kangaroos. With your permission Lt Sharp, I think I am going to move the radio to one of the radar bunkers. Do ya think there might be room? Sure was bouncing back and forth!"

Sharp laughed and in good spirits clapped Frazee on the back. "You just made history and you are working on improving the set up. I will make sure your name is in my report. Thank you very much for proving what my radar can do! Go do what you need to do!"

Frazee sketched a salute for Sharp which was professionally returned. Frazee left while Sharp was beaming with pride in his radar. Dak, smiling, watched Frazee leave then turned toward Sharp, "I'm going to check on my men. Do you have the post?"

"Go ahead. I'm going to call my Sergeants in and let them know what we just did." Dak left and Sharp turned back to the observation slit. He looked at the ROK NCO and four enlisted men, "Hot damn! We just made history!"

Chapter Fourteen

Carpenter was moving up the back trench line toward the command bunker. He hustled a little bit and was wiping his hands as he moved. Just before he turned toward the command bunker, a tornado of explosive debris filled the trench and air. He was knocked to one side of the trench and buried under a rain of dirt and debris.

A few minutes later, a limp body was drug out of a debris infested trench by ROK soldiers. They put it along the trench and the American medic Casper checked him for injuries. He removed his helmet. Then he tended to the body by wiping at it with a rag, cleaning out the mouth and nose. Casper moved briefly aside to let some rescue ROKs by. Slowly, Casper saw the very dirty face of Sgt Carpenter. Blood was coming from his nose and ear. He was the limp form propped up on the side of the trench. Casper poured water over Carpenter's face. Water dribbled into the dirty mouth and the lip quivered. Suddenly Carpenter's chest heaved as he came awake screaming, "Dig me out! DIG ME OUT!"

Casper forcefully pushed Carpenter, the wild eyed Carpenter, back down, yelling trying to get through to him. "You are okay! You're okay. You are safe! Damn it, man! Knock it off!"

Carpenter was gasping for air and uncontrollably shaking. He grabbed hold of Casper with both arms in a death like grip. "You're safe! Relax! Relax. It is over. It's over. You're safe now. Relax!"

Slowly, Carpenter relaxed his death grip on Casper. The physical contact of another living human worked. Carpenter closed his eyes and passed out. The shaking subsided. Casper let him rest against the trench wall. The medic was then startled by another voice behind him.

"How is he, medic Casper?" Lt. Col Kim asked. Casper turned toward Kim and stood up. Casper looked at Carpenter, then back at Kim. Dak headed into the remains of the command bunker.

"IF he has no internal injuries, and no serious head concussion, he should be okay. I am amazed that he has no obvious injuries. That explosion should have killed him."

"A few more feet toward the command trench and he would have been killed. And now I agree with some of my superstitious men. He has the spirit of luck."

"What about those in the Command bunker?" asked Casper.

"All are dead. Lt Sharp, Sgts Quincy and McCoy. We also lost one of my NCOs and 3 enlisted soldiers. They were killed instantly."

Whitehorse, Ransom, Frazee and Cushing arrived by bypassing the trench, going above ground around the workers in the trench, and jumping back down near to Kim and Casper. Ransom squatted near Carpenter, then looked up imploringly at Casper, "Is he dead?"

"No. He is unconscious. I do not know how long he will be out. Keep a light near him. When he wakes up he may panic if he is in the dark."

Whitehorse and Ransom picked Carpenter up and carried him below. The others followed, then stopped, looking at the ROKs bringing out the covered bodies killed in the command bunker. Dak came out holding a helmet.

Frazee said, "Hells, bells. I was just in there. It must have been one hell-of-a lucky enemy shell." Then very quietly in a flash of understanding, "Or an 8 inch." His voice returned to normal, "It went inside before it blew up. I assume the radios are toasted?"

"Yes. All gone. It is ... It is a very bloody place. Not much to salvage. It will be difficult to rebuild."

Dak handed the helmet to Kim, and turned back to the recovery team. The helmet was Lt. Sharp's helmet with the officer's bar on it. Kim took it with his fingers inside. He grimaced and withdrew his hand which was covered now by yellowish lumps and blood.

Frazee yelled, "Cushing! Go salvage whatever antenna wire you can and the antenna if you can find it. I am going to get our last radio from below."

Cushing and Frazee moved off and Casper handed Kim the

rag he was using. Kim then separated the liner from the metal helmet. The liner was cracked and bloody. Kim set both aside and bent down and picked up Carpenter's helmet. He looked at it reflectively for a moment, then separated the liner from the metal shell. Carpenter's liner was dropped into Sharp's old metal shell. Kim used his fist to seat it. He carried it off leaving Sharp's bloody broken liner next to Carpenter's steel metal pot.

Down below, a crack in the water cistern went unnoticed and the water began leaking away.

IX CORP - ARMY HQ
Lt. Gen. Reuben E. Jenkins Commanding
1015 Hours

The scene was almost hectic. Men were moving back and forth urgently, but with a purpose. An aide with a message moved toward a group of three Generals.

"General Jenkins, S2 reported for immediate action." The aide gave the message to Jenkins. Jenkins opened it and stepped to the map table. "Where the hell is Hill 433? Some poor bastards are reported lost up there. Who or what is Bluefin 4?"

The other two Generals came over to look. Major General, Arthur G. Trudeau, 7th Infantry Division commander pointed it out. Stabbed his finger on it.

"Here it is. Hill 433. It had been held by the ROK Capital Division until they got mauled. Now it is behind enemy lines."

"Thanks Arthur. Eugene, Is your 65th Infantry Regiment in blocking positions secure?"

Major General Eugene W. Ridings, 3rd Infantry Division commander replied, "Yes, sir. They have repelled only minor attacks so far."

151

An aide spoke up, "Sir, Bluefin 4 is MIA. Their officer was evacuated. They were an Air Force FAC team assigned to the ROK 26th Infantry Regiment, 2nd Battalion."

"Update your records son. Bluefin is on Hill 433 with a bunch of Americans and ROKS.

Another aide walked up, "General Jenkins, Message from 8th Army command, Army Security Agency." Jenkins opened it and paused passing the message around to the other Generals.

"The ASA left a Top Secret Mission stay on Hill 433? Incredible," replied General Trudeau, raising his voice loudly.

Jenkins thundered to the room, "Listen up! We got friendly forces and VIPs behind enemy lines. Hill 433 is besieged! Get the word out. Designate that position as friendly. DO IT! Someone get Bluefin 4 on the radio!"

"Sir, I can try to send up the first battalion of the 65th. They may be able to push through to Hill 433," said General Ridings.

"Get it started ASAP, Eugene. Arthur, this message says this TS mission was under your wing. What do you know about it?"

"I will have to check on it. I'll be back ASAP, sir!" Trudeau and Ridings left. Jenkins looked at the map and shook his head sadly.

Hill 433
Left Radar Bunker

USAF Sgt Frazee and his side kick Airman Cushing were trying to get their old beat up radio working. All they heard was static. Cushing reached behind the set and jiggled a connection. The static paused.

"That did it. Try it now," Cushing says.

"Mayday, Mayday, Mayday! This is Bluefin 4. Over," Frazee said into the radio microphone. Nothing but a slight static was heard.

"Let's try that old reserve frequency for IX Corp," said Cushing.

"After some adjustments, Frazee repeated his call, "Mayday, Mayday, Mayday! This is Bluefin 4. Over." Both men were startled when a clear voice came out of the ether.

"Bluefin 4. This is Riser 42. We copy your Mayday. Standby for message. Over."

"Roger Riser 42. Standing by."

"Cushing! Go get Col Kim. We got radio contact!" Frazee yells. Cushing ran out. There was a bit of static then another voice came over the radio.

"Bluefin 4. Confirm the opposite of your 6. Over."

"South. I repeat, South. Over."

"Roger Bluefin. Who the hell is in charge? Over."

About this time, Kim and Cushing came in. Frazee looked at Kim and Kim nodded his head. Sgt Park followed carrying an extra helmet.

"Lt. Col. Kim. ROK 26th Regiment. Over."

"Who is in charge of the detachment? Over."

Kim smiled. He looked at Cushing. Parks fingered the helmet he was holding in his hands. He smiled, nodding his head. "Airman Cushing. Please go get Sgt Carpenter. Tell him he has a date with destiny." Confused, Cushing nodded his head and took off again.

"Roger. Many KIA. Acting Sergeant in route. Over."

"Roger that. Standing by."

Carpenter was rushed in by Cushing. Carpenter was bare headed and his eyes were bloodshot; he looked very tired. Kim smiled at him and pointed to the radio. Frazee handed him the mic. Carpenter looked puzzled and took the mic. "This is Sgt Carpenter. Over."

"Sgt. Carpenter, are you the ranking member of your detachment? Over?" Carpenter looked irritated. Like a game was being played on him. He was suspicious.

"Roger. Who in the hell are you?" There was a brief laugh over the radio.

"This is General Jenkins, son. Over."

"Oh, shit. I just cursed at a General. Who the hell is he?"

153

"Major General Jenkins is in charge of IX Corp," said Kim, smiling.

"Oh, just great!" Carpenter turned back to the radio microphone, "Aah, sorry sir. How can I help you sir? Over."

"Listen carefully, Lt. Carpenter. You are now in charge of the detachment. Do you copy over?"

"Roger that sir. In charge. Over."

"It is important that the 'detachment' does not surrender. Do not break out. Stay put. Do you understand, son? Over?"

"Roger. Stay put. Protect or destroy 'detachment.' Over."

"Good son. When this is over, we'll have a talk. Now, put Col Kim on. Over."

Carpenter blew out a sigh of relief. He handed the mic to Kim. Kim addressed the radio, "This is Lt. Col. Kim. Over."

"Col. Kim, Weather report and status. Over."

"Weather is completely socked in. We will hold. Over."

"I'll assume you understand. Do the honors for me, would you please Col Kim? Over."

"Roger, Sir. My pleasure. Over."

"Excellent. I look forward to your report. Monitor this channel. We will tie in assistance. Confirm? Over."

"Roger. Confirm. Over."

"This is ah, Riser 42 out."

Kim looked at Frazee who took the mic back. Kim looked at the tired, beaten up Carpenter who was now leaning against the sandbags.

"Sgt Carpenter, stand up! Front and center!" Startled, Carpenter stood up and took a step forward.

Carpenter saw Frazee and Cushing looking at him smiling, he hadn't figured it out yet. "What the hell? Uh, Sir."

Kim smiled, enjoying Carpenter's confusion. "Attention to orders. Per radio orders, 8th Army, IX Corp command, this date. Sgt. Richard Carpenter is promoted to Second Lieutenant, effective immediately!"

Carpenter was dumbfounded. He stood there as Kim pinned

a brown butter bar to his collar. Sgt Park stepped forward smartly and presented Sharp's old helmet with Carpenter's liner in it. Kim took it and gently placed it on Carpenter's head. Smiling he reached out and took Carpenter's hand, shaking it.

"Congratulations, Lt. Carpenter. You are now an officer." Carpenter jaw dropped as he stared down at Kim shaking his hand. Kim, let his hand go. Carpenter numbly saluted Kim who returned it. Carpenter turned toward Frazee and saw him, Cushing, Sgt. Park, Sgt Cho, and his old friend Whitehorse. They all saluted him with grins on their faces.

"Oh, shit," Carpenter said softly. Numbly, he returned the salute, then the men crowded in, offering personal salutations.

For some reason our Lieutenant wanted to have a meeting of all the NCO's in our unit. It was quiet, and at the appointed time we all started for the Command Bunker. All of the NCO's, and several acting NCO's made it to the bunker. I was held up for a few minutes repairing a machine gun. I knew I was late, and started to hurry toward the Command Bunker. I was about 50 feet away when we got hit by a very accurate Artillery Barrage that bracketed our position. The Bunker received a direct hit, and everyone in it was killed instantly. The explosion blew me back down the trench line, and covered me with debris. Koreans in a nearby firing position saw what happened and immediately dug me out. I was in shock, but otherwise did not have a scratch on me. It took me awhile to realize what had happened. Fortunately the FAC Team, and all the Korean Officers had gone to the kitchen to eat, so they survived. But we lost all of our Officers and NCO's, a total of fourteen Americans and twenty or more Koreans. All of the radio equipment was ruined, and we lost contact with our Chain of Command.

Richard Carpenter, Memoirs

Lt. Col Kim's Quarters

"Lt. Carpenter?"

"Yes, sir?"

"There is something that you need to know. Many leaders are not born leaders. They are often common men such as yourself, forced into leadership when the situation dictates."

"I agree with that Colonel."

Kim nodded his head briefly, then chose his words. "Leaders are made. Leaders often learn the hard way, they make mistakes. The learning curve is high. And when WE, military leaders, make mistakes, men will die. Know this, we will make mistakes as leaders and we better damn well learn from them."

Kim reaches out and draped his arm over Carpenter shoulder. Then with close eye to eye contact, Kim drive in his final point. "Together. As a team. We will make less mistakes. Do you understand?" Carpenter nodded his head as Kim pat his back.

1145 hours

The shell shocked Cpl Andy Livingston was crying once again in the corner of his bunk. He whimpered, "I want to go home." Livingston got up and unnoticed, walked up the ramp and then numbly started toward the airfield. The ROKs at the edge of the airfield were stunned seeing an American walking without a weapon, helmet or gear and exposed above the ground. Two ROKs looked at each other and one ROK soldier made the crazy gesture with his fingers. Sgt. Myung-ki Cho ran up to them, "Who is that crazy fool?"

"It is that American who put his friend out of his misery when he was burning."

Cho looked up at the sky, "Why me?" Sgt Cho jumped out of the trench and ran after Livingston. The two ROK soldiers looked at each other and one shrugged his shoulders and sighed. They jumped up and protected Sgt Cho as he brought back the crazy American. The one ROK soldier (Ok Joe) shook his head as Livingston muttered, "Where is the bus stop? I am lost."

"Joe. Not. OK."

In Col Kim's office, Sgt Cho was reporting what he saw the crazy American do. Cpl Livingston was standing between two ROK soldiers muttering about going home. "And he has been muttering about going home, being lost and something about a bus, sir."

"But, he was walking away without gear?"

"Yes, sir. He was walking away, but not trying to hide."

"Either he was trying to surrender or desert. This is most unfortunate."

Lt. Carpenter knocked on the open door. "Sir. May I ask what is going on?"

"Your Livingston was deserting us."

"Shell shock. A psych case."

"We do not have such luxuries here, Lt. Carpenter. The man is a danger to all of us and a deserter. He must be shot."

For a moment Carpenter was stunned. Then he turned in place, staring at Livingston. He stepped in front of Livingston. "Where were you going?"

Like a child, he responded, "To the bus. I am late going home. I lost my fare. May I please have some bus fare?"

Carpenter looked at Livingston sadly. Carpenter turned awayfrom Livingston. He looked at Sgt Cho and the two ROK Soldiers. "Sgt Cho, please take this wreck of a man to his bunk? And have the medic Casper look at him. Guard him closely, please."

Cho looked at Kim. Kim nodded. Cho left and put the privacy curtain in place. "That man is a deserter. Justice must be served."

Carpenter took a deep breath and turned to face Kim now that they are alone in the room. "Colonel, that man is sick. Shell shocked and not in his right mind. I will not order him shot."

Kim looked angry and leaned forward toward Carpenter. "That man walked away from this command. At best the Communists would just shoot him. At worst they would have questioned him and found out about your precious radar and details of this position. Shell shocked or not, that man is a liability. He is no good to me, to you or anyone. The best he can be is an example of what happens to a deserter. You need to shoot him."

Carpenter looked at Kim, a bit unsure of how to proceed, but getting angry at Kim. He looked down at the ground trying hard to figure out a way to save Livingston. Kim saw the hesitation and misunderstanding, reached out to touch Carpenter's shoulder. "I understand. You are a new officer and an American who is not used to such decisions. I know it is a hard thing. That is why I offer my services to do it painlessly. It is best for all concerned."

Carpenter looked up and stated shakily. "No Sir. I will carry out any order you give, except this one. That man is my responsibility." He then said firmly, "And I will not let him be killed as a deserter." Kim's face went red, but he controlled himself. After a moment, very coldly, he addressed Carpenter.

"Then he is your responsibility. If he 'leaves' again, then I will have your head, Lt Carpenter."

Carpenter came to attention and saluted with a crisp, "Yes, Sir. He is my responsibility." Carpenter brought down his hands smartly, then did a perfect about face and exited the room. As the privacy curtain settled behind him, Carpenter muttered, "Shit and double shit."

Chapter Fifteen

Radar Bunker

Sgt Park entered and motioned to Carpenter to come outside. Carpenter followed. "Lt. Carpenter, my men report that Corporal Riney and his men are preparing to leave. Did you order this?"

"No fucking way. Where is that bastard Riney now?"

"He and most of his men are in the American quarters." Carpenter headed for the bunker downstairs. He looked angry. As he got to the door of the American bunk room, he saw men packing. He stepped into the room.

"Riney, What the hell are you doing?"

"None of your business Carpenter. I don't care about your fake rank…" Riney tried to bully his way past Carpenter. Carpenter grabbed Riney's arm and used Riney's own momentum plus the force Carpenter exerted to force him into a spin. When Riney was facing the way he came, Carpenter let go and gave a slight push. Riney hit the bunk hard and now was in a fighting mood. He jumped to his feet and rushed Carpenter. Carpenter was in a martial arts ready stance. Carpenter twisted left and did a right side kick into Riney's chest. With a large "whoof" Riney fell backward onto the ground.

Carpenter slowly pulled out his 45 and checked it to make sure a round was in the chamber. In the dead quiet, he heard the safety "snick" off. Riney's eyes went wide staring at the gaping circle of death in front of him being pointed at his

Face.

"If you are a danger to the detachment and this position, then you should be shot as an example to others." Carpenter looked at his pistol ,then put his weapon away and took a deep breath. He looked at Riney with gritted teeth. "I need leaders. Not dumb shits. I need you to be a leader of men. I need you to be the squad leader of 2nd squad. Can you do this Sgt Riney?"

Riney looked relieved, but stunned. This was not the Carpenter he had known. Riney coughed then nodded his head. Carpenter reached down. Riney looked at, then took, Carpenter's hand. Carpenter then pulled Riney to his feet.

They looked each other in the eye and understanding came.

"Sgt Riney. Gather all the Americans down here. We need to talk."

"Yes, sir."

Bunker Common Room

The surviving Americans gathered around. Carpenter looked at them as an officer. He knew that what he would say would not be popular, but needed. Even some ROKs watched.

"As you all know, Yes, we are surrounded by a million or so of the Chinese Army. Yes, we rejected their offer to surrender. Yes, things do look bleak. He paused. But let me tell you something. We will stay here because we have orders from General Jenkins to stay put. We are not to break out or surrender. Those are our orders. ... We have our amazing White Tiger brothers here willing and able to fight. We have Col Kim. We have Sgt Park. We have Sgt Frazee and we have our radar." He took a deep breath.

"And if you don't like it, tough. It is time to prove we are soldiers." Carpenter now firmed his voice, "I intend to go home to my family. And I ... do not ...give a shit ...how many ... of the enemy ... I need to kill to do that. To go home, I need your help. That way we all can go home. ... Now, let's go get ready."

"First squad! On me! Let's do it," Whitehorse commanded.

"Second squad! On me! Move out!" said Riney.

The Americans turned and hustled out and up the ramp. They followed Sgts Whitehorse and Riney. The ROKs even hustled out.

Kim prepared to blow up the supply dump, while Carpenter readied white phosphorous grenades to destroy the radar sets if they were overrun.

"Do not, under any circumstances, allow those Commie bastards to get this radar. Just remember to get the hell out of the way before it blows," he said.

The radar specialists nodded their understanding.

Acting Squad Sergeant Whitehorse was just finishing writing a final letter home. Specialist Rodriguez came up to him holding a folded sheet of paper. Rodriguez coughed and seemed reluctant to talk. Whitehorse glanced up at Rodriguez, seeing the paper in the man's hands. As he talked, he indicated his letter.

"A letter home? It's probably similar to mine."

Rodriguez nods his head and softly says, "Will you hold this for me? Just in case?"

Whitehorse took the letter. Rodriguez, a Catholic, nodded his head then made the sign of the cross. "Thank you. I just got that feeling. It is not a good

feeling. I just hope I don't screw up."

Whitehorse nodded taking the letter with his left hand. With his right, he held his out his hand to Rodriguez who shook it briefly. "I know you. You won't. You're a good man."

Rodriguez then turned away with a tear in his eye and walked out.

Whitehorse folded his letter into thirds. He placed Rodriguez's letter on top of his. Then he reached down and picked up three more letters he had been given. Very carefully he placed them in small dispatch pouch. He then stood and walked down the trench.

Lt. Carpenter, while writing a letter to his wife, saw Whitehorse come up then wrote some more. After writing a bit he realized his friend Whitehorse was being quiet and standing a few feet away until Carpenter was finished writing. Carpenter smiled at his friend, "What do you have Sgt Whitehorse?"

Whitehorse seemed oddly reluctant to talk then handed Carpenter the small pouch. "I got bad feeling."

"Bullshit. We are going to get through this. I am going to see my family again. And you are going to marry that girl you keep talking about."

"Maybe."

"No ifs, ands or buts." Carpenter stood up and reached out to his friend.

"Keep it in a safe place anyway. Please."

Reluctantly Carpenter accepted the pouch. Whitehorse then reached his hand out to Carpenter. Carpenter took it and did not let go. Still holding Whitehorse's hand he looked the man in the eye. "We are going to give them so much hell they will never mess with us again."

Whitehorse smiled weakly. "Yeah. We, Lost Bastards will give them one hell of fight."

Whitehorse then left and passed Acting Squad Sergeant Riney who was heading to see Carpenter. Riney was carrying a small dispatch pouch. The two men nodded neighborly at each other in passing. As Whitehorse worked his way back up the trench, men were sharpening knives, shovels, cleaning weapons and doing some

Praying. Livingston, the American who'd had to kill his friend who was burning to death, was seen helping the Casper, the medic. "How are you holding up?" Casper asked. Livingston sighed and helped Casper bandage the wounded men.

"Doing penance. I am beginning to remember. Everything."

"Saving lives does help one's soul."

The men went back to helping the wounded. There were a lot of them in the makeshift hospital in the bunker.

Radar Bunker 1

Carpenter walked into the radar bunker. Rodriguez and Larry Smith, known as "Smittee," were looking at the radar screen.

"Lt. Carpenter, those bastards are lined up and ready to the north about 700 yards out. Smittie, have the ROK turn to the west," commanded Rodriguez.

Smittee went out and the screen went nuts for a moment, then settled down.

"You can see part of them lined up to the west. It looks like it will be a two sided attack."

"Crap, this is going to be difficult."

Smittie came back inside and they heard the Chinese bugles begin to sound. The enemy to their north began to advance. Artillery began to hit off to the west, smacking the Chinese lining up over there. The Chinese ran for cover.

"Bless Frazee and his minions! The attack to the west looks like it has been broken up over there. Smittie, have the ROK turn back north," Rodriguez directs once again. Carpenter smiled at Smittie as he left once again.

"You keep an eye out for the western attacking force. If they reform or come at us, let me know PDQ!"

"Roger that, Lt. Carpenter." Carpenter shook his head. The radar screen went nuts for a little bit before settling down.

"That still sounds strange to me." Rodriguez chuckles a bit.

"Enemy now approaching 600 yards," says Rodriguez. The sounds of heavy artillery rumbled overhead. Explosions were heard.

"Crap. Too much debris being flung up. I can't get a reading." The power phone rang. Rodriguez grabbed it quickly.

"Radar 1. ... No, my screen is scrambled also. I think they were short, though.

"Tell Frazee to request a star shell barrage," Carpenter directed. Carpenter then headed out past Smittie coming in. He scrambled up the sandbags next to the ROK soldier manning the radar. He pulled out his binoculars and looked through the misty darkness to the north. Carpenter panned the cloudy and smoky hillside. Artillery rumbled overhead. Suddenly there were bright flashes like fireworks in the clouds. Then one by one, six star shells began to float downward through the misty clouds.

In the distance a lone Chinese bugle blew, then the mournful tune was picked up by others. Incoming enemy Artillery hit out in front of the defensive position. The enemy artillery started walking up the hill toward the defenders. Suddenly a long line of Chinese infantry appeared out of the cloudy, smoky mist, striding up the hillside. Carpenter, using his binoculars, panned across and it appeared that the Chinese line was a wall of men as far as he could see. "Oh, shit!"

Chinese whistles blew and the enemy began to charge early. After a pregnant pause, a commanding voice yelled, "Open Fire!" Defensive machine gun fire went out and the battle was on.

Tracers reached out and touched the enemy. It appeared that each tracer touch knocked down dozens. But the enemy was relentless and kept on coming despite the losses. Other star shells lit up the scene with their magnesium light casting shadows as they rocked back and forth under their parachutes. The defenders of Hill 433 began to see and hear individual rifle fire, then incoming enemy artillery caused the allied soldiers to duck.

The attack had begun to the north while the western Chinese attack seemed foiled. Artillery rounds went out and pounded the enemy to the front. ROKs moved toward that area of attack. This time instead of retreating, the enemy took cover and held their ground while they continued to shoot. Small groups rushed forward here and there. The enemy hid amidst the dead and dying. They got pounded by the American artillery.

Carpenter looked over his shoulder toward the west. He used his binoculars but did not see anything. He tapped the ROK soldier on the shoulder and pointed west. The ROK turned the radar that direction and panned back and forth. Carpenter slid down the sandbags into the trench. He went into the radar bunker.

"Rodriguez, has the enemy to the west reformed?"

"Just a few groups reforming. Maybe they are reluctant."

"They are taking too long to reform. And those bastards to the north aren't pressing in hard. Something is not right." Smittie pointed toward the bottom of the radar screen.

What is this? It looks like water waves."

"That is a strange return. It ripples," said Rodriguez.

"Get on the phone and get me some illumination to the west!" ordered Carpenter.

Carpenter went out the other side of the bunker. There was an artillery explosion in the middle of a trench. The duckboards were tossed and briefly communications lines were seen, as they too were blown asunder. Carpenter moved down the trench to the 50 caliber machine gun position, past Ransom who followed him out the other side.

In the clouds above him to the west, pops of white started burning off clouds.

Suddenly two popped out together, and under the glaring white light, parachutes swinging back and forth, the enemy first and second waves were revealed crawling forward on their stomachs. The ground appeared alive with hundreds if not thousands of humans coming right at them from the west.

"Oh, my God! Hail Mary, full of Grace …"

Carpenter turned and jumped inside for the power phone. He put it to his ear then frantically twisted the crank. He listened again then slammed the phone down

"Shit! The phone is dead!"

Carpenter pointed to Chin Ho Sun, one of the ROKs in the

Bunker. "Sun! Runner! Alert! Chinks to west! Pass it on. Do you understand?"

The young Chin Ho focuses intently on Carpenter, nodding his head at the end, "Yes sir. Alert. Chinks coming west! Tell all!"

"Good. Go man, go!" The young ROK soldier raced down the

trench toward the east to warn others of the sneak attack.

Carpenter turned back west where a few of the defensive machine guns began to fire. The star shells began to sputter out. Carpenter saw Ransom turn toward him.

"Ransom! Gather everyone you can. Block this trench. Don't let them by you!" Carpenter then started moving westward down the trench.

"Yes, Sir? But what …" Carpenter turned and interrupted Ransom.

"Shoot off as many flares as you can from here. You hold them, understand?"

Ransom nodded and Carpenter rushed down the trench. More firing was heard. Chinese whistles were heard. A flare opened up overhead. The Chinese began to rise from the ground by the hundreds and moved forward in mass. The two human waves of men about 50 or so yards apart seemed unstoppable.

As the first wave approached the defense perimeter, mines, improvised explosives, and grenade trip wires hardly slowed them down. A supporting 30 caliber machine gun jammed. The gunner frantically tried to clear the gun, as his position was overrun by a swarm of Chinese firing burp guns down into the emplacement. They took no prisoners. They surrounded the far left 50 caliber machine gun nest. A half dozen enemy grenades were tossed into the bunker. It became a free for all with close combat as some of the enemy bypassed armed defenses and tried to overrun the rest of the positions. Men fought and died where they stood. Some grappled the enemy in mutual death. The battle had now devolved into small local fights.

A group of the enemy suddenly popped up between Ransom's defensive position and the radar bunker. Smittie appeared with a rifle ,firing repeatedly until he was cut down falling forward. Many were killed, but several ran into the bunker, still firing their weapons. Radar specialist Rodriguez was shot and mortally wounded. He looked at the radar and reached over and to pull the wire holding the pins of the white phosphorous grenades to destroy the radar. He killed himself and many of the enemy. The white hot burning phosphorus gave an eerie glow to the battle, while burning the inside of the bunker.

Ransom was firing his BAR rifle as the ROKs cleared a machine gun jam. As his weapon clicked open after the last round, he grabbed a grenade, pulling the pin and tossing it. He grabbed another grenade and tossed this one after allowing it to hiss for a second or two. He then ducked and grabbed another magazine. Somehow he fumbled the reload and the magazine did not fully seat. He allowed the BAR bolt to go forward. He saw an enemy aiming at him and raised the weapon but the weapon did not fire. The enemy soldier fired and Ransom jerked from the impact. Very carefully, Ransom slapped the bottom of the BAR magazine and cycled the bolt once again. He saw a round chamber. He smiled and fired at the enemy now above him. Two rounds knocked the enemy backward. With some effort, Ransom got the BAR into position on the edge of the trench, then began to fire each round very carefully in slow measured fire, shot after shot. That smile was still on his lips as he shot, when debris fell on the now un-responding face of Ransom who had died.

Carpenter fired his pistol at the enemy, knocking several down. When it emptied, one enemy soldiers jumped at him in the trench. Given a moment, he grabbed his knife and jabbed at one man like a dagger. It pierced the man in the neck. Carpenter struck twice more through the enemy's hands which were now down trying to stifle the massive flow of blood.

Carpenter jumped up from the trench taking the leg out from under an enemy soldier. He pulled the man into the trench, stabbing him to death. He took the enemy burp gun away and coming up sprayed a group of the Chinese. He started to retreat back down the trench line. An enemy grenade hit him in the chest, then fell into the mud at his feet. He stomped on it then took a quick step backward onto the duckboard (a wooden walkway allowing walking over the mud). When the grenade went off the duckboard rose but the erupting mass of mud and layers of board protected Carpenter from the devastation of the blast. The heaving duckboard tossed Carpenter backward out of the trench.

A bit stunned, and now sitting above the trench, Carpenter saw three more enemy approaching. He used the burp gun to kill two of them. At close range he fired at the last one standing. Somehow he missed and his weapon ran dry. The enemy soldier aimed his weapon carefully at Carpenter who was watching stunned.

As the man fired his weapon, Carpenter was suddenly pulled into the trench. The enemy soldier did not see the grenade Whitehorse then tossed gently over the edge of the trench. The grenade went off, killing the enemy soldier.

Carpenter looked up at Whitehorse with surprise. Whitehorse grinned at him as he handed Carpenter an empty 45 caliber semi-auto pistol. Carpenter reloaded the pistol shakily, and he began to smile at Whitehorse while he struggled to his feet.

Two of the ROKs with Whitehorse were killed by a sudden burst of gunfire and Whitehorse emptied his M1 clip into the Chinese soldier who did the deed, then reloaded. Carpenter fired a few shots then, when his friend had reloaded, he grabbed one of the ROK rifles. He checked it, then pointed it down the trench. Suddenly there was quiet.

Carpenter and Whitehorse moved to stand back to back in the trench.

"Chester. You know, I think I know how…" starts Carpenter.

"Bullshit General! Be ready!"

The men began to slow fire as the enemy appeared. They had no place to go and nothing they could do but to make their last stand together, back to back.

Private Chin Ho Sun had just reported the situation to Lt. Col Kim. Kim suddenly strode off quickly down the ramp into the bunker. Sun followed armed with an M1 Garand rifle at port arms. It was empty with the bolt locked open. He was holding a full clip of ammo. Kim went down to the hospital and looked around at the wounded men.

"Sometimes I must ask the impossible. Forgive me. I need your help now. All who are willing, follow me."

Without a look back, Kim turned and quickly walked out. Sun

looked down at his weapon and realized it was unloaded. He saw the full clip in his hand, he loaded the weapon and released the bolt. It clanged shut. Without a look back, he followed his commander. For a moment no one followed, then one wounded soldier followed and picked a weapon and ammo from the ready rack. Then dozens followed. The wounded, some barely able to walk, rose and staggered forward, grabbing weapons. They looked determined.

Livingston saw the wounded rising and following Kim upstairs. He bit his lip and looked at Casper still working on a wounded man. Livingston joined the line, grabbed a weapon and ammo, and joined the wounded in the counter attack.

The wounded ROKS counter-attack hit the Chinese running toward the center of their position. They fought them off in a brutal, no holds barred, toe to toe fight. Kim was seen firing shot after shot into the enemy. He holstered his empty weapon and picked up a discarded one. Kim moved deeper into the fray, with his men following. The ROK martial arts and sense of valor started to turn the momentum.

Col Kim was leading the counter-charge when an explosion knocked him down and he lost his weapon. He saw an enemy soldier appear suddenly from the smoke of combat and tried to draw his handgun. The Chinese soldier fired his automatic weapon. But, the American, Livingston, was suddenly charging in front of Kim firing his weapon at the enemy soldier, who went down. Livingston fell forward onto the man who he just killed and who had just killed him. Kim reloaded his handgun and moved forward to check the American. Livingston was dead, and Kim gently touched the dead man's shoulder. Kneeling, Kim muttered a quick prayer, then returned to the fight.

The counter-attack swept the enemy from the ridge. Kim strode forward and found Carpenter and Whitehorse still firing their weapons from the trench. Whitehorse saw Kim and the other ROKs. He tapped Carpenter on the shoulder to get his attention, and quit firing. Carpenter turned and saw Kim. Carpenter tried to get up and stumbled. Kim grabbed his hand. Carpenter looked as dead tired and beaten up as he felt. In the background, Whitehorse began to collect weapons, reload them, and lay them down ready to use.

"Thank you Colonel."

"I was wrong."

Carpenter was too tired to ask what about. He raised his eyebrows as a question.

"Livingston was a good man."

Carpenter numbly nodded his head. He reached for his canteen, he fumbled with the screw off lid, his hands shaking. Kim took the canteen and opened it for him. Carpenter took a long drink.

"Now, my young friend, it is time to show the men that officers are still in charge. Lt Dak is watching the east and I will check the middle while you see to the west side. We don't have much time. They'll hit us again."

Carpenter nodded his head then started down the trench. Whitehorse started to follow him and Kim held him back. Whitehorse glared at Kim.

"See to your squad, Sgt Whitehorse."

Then Kim looked toward Sgt Park who was watching from nearby.

"Go with Lt Carpenter. Try to keep him alive." Park nodded and headed out after Carpenter. Whitehorse nodded at Kim, then went to check on the remains of his squad.

Fortunately the FAC Team had a spare radio stored in their quarters, and they were able to put it into operation. Because all of the FAC's, SSI's, call signs, and code books were lost in the explosion we lost our air and artillery cover. The FAC sergeant was able to establish communications with an Eighth Army Artillery Communications Center, and we were able to get Artillery Coverage from a Battalion of 8 inch guns that had the range to reach our position. We were instructed to use air panels to mark our position so that friendly air strikes in the area would not aim at us. As quickly as we marked our position we had increased air activity over us, and sustained air attacks on the Chinese surrounding us.

The FAC was able to talk to the aircraft, but without the SSI, call signs, and code books he could not direct their attacks, until our position was marked, and he established voice identification with a pilot he knew. The FAC also acted as our forward observer, and directed the fire of the eight inch battalion supporting us. Without the air and artillery support we received, I do not think we could have held our position as long as we did. The FAC had the only map of the position. He had it all marked up with barrages, and artillery fire concentrations, and could call in very accurate artillery fire when we needed it. I think he had just about everything he could see plotted for firing. Plus he had the coordinates to direct the air strikes. He was a very valuable member of our garrison, and certainly earned his pay.

On the 15th of July I was called to the command bunker. There was a general from Eighth Army calling me. The FAC had identified me as the ranking person in my unit. I identified myself as a corporal, and the general immediately told our unit I was commissioned as a second lieutenant. The general then ordered me to take command, to maintain our position, and not to either surrender, or try a breakout. He said that relief was on the way. I told him that Colonel Kim was in command. And he said yes, of the Korean troops. But you command the Americans. Colonel Kim was receiving the same instructions that I had received., and we should work together to hold out until we were relieved.

Richard Carpenter, Memoirs

Chapter Sixteen

Bunker

FAC Sgt Frazee was yelling at his microphone, "Dammit, I need it now! We don't have another twenty minutes! Blue Fin will be gaffed & gutted PDQ! Over!"

Just then an artillery explosion rocked the bunker. Dust flew everywhere. As it cleared, Frazee was protecting the radio. Others picked themselves off the ground. Frazee tried the radio. "Damn it all to hell and darnation! The radio…"

Cushing jumped up and fiddled with the radio. "It's gotta be the antenna."

Cushing rushed outside. Frazee started checking the antenna line, but even without much of a pull, the line fell from the wall. Cushing came back in gasping, "The aerial is gone. There's no sign of the antenna."

"Holly Molly, what a fucking mess!" Frazee looked around angrily as if looking for something to hit. Suddenly he stopped, then turned and glared at Cushing. "Go get that portable antenna from our quarters."

"That is a short range antenna, almost as good as line of sight."

"I know. Go get it, now flyboy!"

Cushing rushed off and Frazee disconnected the power line and placed a fresh battery in the old radio.

Sgt Park joined Lt. Carpenter on the far west side of the position. Not many defenders were left. Most were wounded and many seemed numb from the hard fighting. Exhausted, they saw the early morning light trying to burn off the fog and clouds.

The power phone rang. Carpenter answered. He listened, then replied, "Roger." He hung up the phone. "The radar reports they are coming at us again from the northwest. Another group is beginning to form up to the north."

Carpenter looked around at the few defending the position. Wounded men were assisting by bringing ammo and grenades up from below. The remaining allied soldiers readied for the attack. A lone Chinese bugle blew, followed by others. As the misty clouds cleared, artillery explosions were seen walking toward the allied soldiers. Then suddenly a long line of Chinese infantry appeared striding up the hillside.

Whistles blew and they started their charge. Defensive machine gun fire went out and the battle was on.

The ROKs and Americans steadied themselves for the attack, in which they did not think they would survive. Carpenter took another quick gander between enemy artillery bursts. "Damn! No artillery is hitting the enemy."

In the trench next to Sgt Park, Carpenter took off his helmet and looked at the picture of his wife and children inside. Slowly he put it back on his head. He gave a brief smile at Park. "Sgt. Park, I have learned a lot from you. And from Colonel

Kim. You are truly professional soldiers. Thank you for what you have taught me and my men." Carpenter stuck out his hand and Park shook the hand.

"I too have learned much. You are a good man Lt Carpenter." Carpenter nodded his head then closed his eyes for a few seconds.

"I really do miss my wife and kids."

The enemy rolling artillery barrage lifted from the western edge of the defensive position. Slowly, Carpenter stood up and stepped up on a box. He looked at the enemy a little over 100 yards away. He looked around and the remaining defenders could see him. They turned to look at him. With a deep breath, Carpenter, the leader, saw that the ROKs and the few Americans were watching him. He screamed out a growl, "For the 26th!!!"

Barely a moment later, Col. Kim was some 75 or so yards away. "White Tigers!!!" The men were startled and energized. With a growl and as one they rose to fight. Carpenter looked at Kim and smiled.

As one they yell in Korean, "Open Fire!!" They all opened fire with the remaining weapons. Carpenter stepped down and picked up a BAR. He started taking quick single shots with it. Sgt Park watched Carpenter as he, too, joined in the fight with his weapon.

The enemy line kept coming closer and closer. The gaps in the line keep getting closed. In the early light the enemy became distinct and clear. They appeared determined. One of the nearby machine gun jammed. The gunner pounded on it in frustration. The enemy was now close enough that they started firing their personal weapons. The defenders began to fall. Grenades flew back and forth in a bizarre game of catch.

Near Radar Bunker 2

Frazee was working the knobs while Cushing finished attaching the small metal antenna. "We can't get any of the artillery batteries because we are on the wrong side of this hill."

"I know. Cross your fingers." Frazee seemed unruffled as he picked up the microphone.

"Mayday, Mayday, Mayday! This is Bluefin 4, declaring an emergency. Over."

The radio crackled and then a voice was heard.

"Bluefin 4, state the nature of your emergency. Over."

"This is Bluefin 4, I'm stuck on a hill about to be over run by a thousand Chinese infantry. And I really, really need some help. Over."

"Bluefin 4, This Whisky Flight 72. Authenticate, over."

"Damn it, I can't! Listen, please! Bluefin is a tuna, knucklehead. It is an open water fish. Ya catch it on a bare hook and gaff it, if ya gotta. Come on boys, I need help here! Over."

"Bluefin 4, If you cannot authenticate…" Sudden static is heard on the radio.

"Bluefin 4. This is Zebra 2 Niner. Over."

"Zebra 2 Niner. Is that you box face? Over."

Zebra 29, Blue Sky Raider aircraft in a flight of 6 aircraft, replied, "Roger that! All flights Bluefin 4 is authenticated. He needs help! Rally up boys. ... Bluefin 4, what is your location? Over."

"I am on Hill 433 about 3 miles south of where you visited. We are on the north military side. We are getting enemy artillery fire. Over."

From the sky several puffs of smoke rose from 3 different areas. Bluefin 4, I see 3 strikes in progress. Can you provide more details? Over."

"The artillery just stopped. That means they are close. Over."

In two areas the smoke and light flashes continued, but one started to drift down wind. The flight headed directly for that hill to the west. "I see ya Bluefin 4! I am coming from your east with a flight of 6. We have jelly but no donuts. Standby."

Zebra 29 changed to his flight's frequency. He called out to the other members of his flight, "Zebra flight. Okay boys, Bluefin 4 is one of us. We break into two flights. Those on me echelon right. Then every one change to the guard channel. Do it now."

Three aircraft switched to a stepped right formation where the lead aircraft was in the lead with the next aircraft slightly to the right rear and the third to the right rear of the second. Zebra flights 1 & 2 switched to the guard channel.

"Zebra flight 1. Watch me. Pickle when I do. Copy?" Two quick double-clicks of their mics relayed 'copy that' from the other two pilots.

"Blue Fin popping red smoke! Can you see it?"

"Bluefin 4. Zebra 2 Niner descending with 3. I see red smoke. Over."

"Zebra 2 Niner. Bluefin 4. 150 yards west north west. Over!"

"I see enemy in the open! Standby, Bombs away!" The large containers dropped away from the underside of the 3 aircraft. They tumbled over and over the sandbagged trenches below. Tracers streamed outwards from the bunker. It appeared the containers were going to hit the sandbagged trenches. But at the last fraction of a second, the containers hit and for a second they drenched the advancing enemy soldiers with a clear liquid. Then with a swollen whomp, it flamed into burning napalm. Suddenly and without warning napalm hit the Chinese just yards from the front of the defenders. The defenders ducked from the severe heat. The ROKs closest to the napalm ducked as flames washed over the top of their trench.

"White Stars! White Stars! White Stars!"

Carpenter turned toward Park as they both ducked from the heat of the napalm. He had a maniacal grin. "Frazee did it! He really really did it!"

The next flight dropped their "eggs" on the fleeing Chinese. Napalm caught them running for their lives.

The Chinese next attacked us from across the top of the ridge line to our south. That was an early morning attack that our radar missed because of the ridge. Fortunately our massed fire, plus the artillery support, and air cover, stopped the attack short of our position. None of the Chinese even got close. One thing that stands out in my mind about that attack was our planes dropping napalm in areas just in front of our positions. They would come in so low, and so fast that it was difficult to see them through our firing ports, but we sure felt the heat of the napalm.

On the 16th of July the Chinese tried to overrun us one last time. This time it was another night attack, and they came at us from three sides, east, west, and south. We called in the prearranged artillery fire, and aircraft dropped parachute flares for us to see. It was a close call, and two Chinese were actually able to get on top of the trench complex. Colonel Kim saw them, and killed them both with his pistol. Near midnight the attack fizzled out.

Richard Carpenter, Memoirs

Chapter Seventeen

Carpenter and Park stood and surveyed the devastation of man-made hell on earth. The earth was blackened and smoky. Men placed air-recognition panels on the edge of their positions. Aircraft flew over and waggled their wings. The wind shifted and smoke wafted by Carpenter and Park. Carpenter's nose wrinkled then he gagged from the smell. He retched from the stench of burned flesh and hair wafting in on the wind. Sgt Park handed Carpenter a canteen to rinse his mouth.

"Breathe through your mouth."

Carpenter nodded but still looked like he would be sick. They ducked back in the trench when a sniper shot at them. Then, as several aircraft made a strafing pass they watched a single engine aircraft with two men in it fly over. As they came back over, they airdropped several packages into the position. The packages floated down by small parachutes. Carpenter and Park ran down the trench to where the ROKs were bringing in the packages. They started to climb out of the trench. Suddenly there was a crowd of ROKs climbing out of the trench with them. Carpenter looked around and realized they are a target.

"Scatter. Go!"

The men quickly obeyed and scattered. Carpenter jumped back into the trench. As men recovering the packages were almost to the trench, long range shots came in, hitting one of the ROK soldiers in the leg. He fell, tossing the package to another soldier in the trench. The man was almost within arm's reach. The man looked at Carpenter. Carpenter jumped up and out snagging the man's arm. As Carpenter pulled the man in, Park pulled Carpenter back in too. As the man came into the trench more sniper rounds hit where Carpenter had been.

The men looked at each other and started to laugh. Carpenter laughed too.

"Damn that was a stupid thing to do."

"Your friend Whitehorse, would not be pleased if I allowed you to be shot."

Carpenter nodded his head soberly as they followed the ROKs down the trench to the ramp downstairs. They saw Frazee kneeling using a pocket knife to open the packages.

The first package was wrapped in layers of bandages. Frazee handed it to one of the ROKs. "For medic Casper. Antibiotics and bandages."

Frazee tapped the second package. It was solid. With a quick swipe of the knife he revealed batteries. The larger package had a lot of cardboard around it with rags in between. He smiled as he slowly cut it open. It was a brand new radio. Frazee laughed as he removed three cigars from where they were protecting the knobs. Then he pushed both packages to Cushing laughing. He sniffed the cigars. "Come to Papa!"

Frazee smiled then shook his head. He looked up and saw three ROKs. The middle one who had been shot was being held up by the others. With a grin, he stood up and presented each one with a cigar. For the middle one, he put the cigar in his mouth then ruffled the man's short hair. They smiled.

"Go see the medic!"

The ROKs left as Frazee turned towards Carpenter. "I wonder if Cushing has the new radio warmed up yet? I got a bone to pick with the people who were recently trying to kill me. If you would excuse me gentlemen?"

Carpenter nodded his head, and watched Frazee whistle as he headed up the ramp.

In radar bunker 2, Frazee and Cushing called in artillery and airstrikes around Hill 433. They were calm, business like.

There were many men still in the trenches. They were tired, worn out. Many were wounded. They were almost listless. Some had makeshift scarves around their faces. Occasionally, rifle shots echoed off in the distance. As one ROK looked out the firing slit, enemy MG fire suddenly hit the bunker. One bullet hit the firing slit and ricocheted inside. The ROK flinched and raised his hand to his face, where there was a small amount of blood. "It cut me."

180

"Well it bit me," said another ROK soldier, who had a wound in his upper arm. He grabbed it and held pressure as blood seeped out. Whitehorse looked at it. And smiled at the man. "It is not bad. Go see the medic."

"Yes, Sgt." The man left the bunker and Whitehorse turned toward the other ROK Soldier who was using his canteen.

"That is the last of your water. Go easy." The ROK soldier nodded his head, took a brief sip then continued to watch for the enemy.

Evening
Hill 433
Near the Command Bunker

Carpenter was looking up at the stars that evening from inside the communications trench. He was pensive. Sgt Frazee joined him.

"A penny for your thoughts," said Frazee.

"I was thinking about all of this. My goal all along has been to do my job so I can go home to my family. That is my focus. But ... the peace talks have continued for over a year. Three months ago they exchanged the sick and ill. But still the Chinese attack like they do not want peace."

"It is all politics. The Chinks understand that war is another form of politics. Life is cheap for them."

Frazee paused and looked up at the stars for a moment. "In war we often have to be as cunning and as cruel as the enemy. But what happens when one day we wake up and find ourselves worse than our enemies?"

"Shit plus three still makes shit. Then we have to clean it up," replied Carpenter.

"You're right. War stinks like shit. Then we have to clean up the mess."

A ROK runner came to Carpenter with a message. He looked at it, then handed it to Frazee. "Time to see Colonel Kim." The two Americans headed downstairs to Col Kim's office. He was not there. They turned and headed deeper into the bunker. They saw Sgts Whitehorse and Riney walking up from below with Col Kim.

"Lt Dak was hit by a sniper. It is a chest wound," said Kim.

"Then this may be good news. Radio message reports possible relief sometime tomorrow morning by standard route," replied Carpenter.

"That is indeed good news." Kim turned around and headed deeper into the bunker leaving the Americans alone.

"You know, it would look good if we had a flag flying to greet them," said Riney. Whitehorse grunts then walks off.

"Well, in the meantime, we put out more air recognition panels. And notify everyone to expect a relief column up the road," directed Carpenter.

Whitehorse suddenly appeared unfolding a small American flag. He held it open and all the Americans touched it. Several had a visceral emotion toward the symbol. Carpenter wiped a sudden tear away. "Sgt Whitehorse. I don't know where you got it. But would you do us the honor of mounting and flying it?"

"Yes, sir. It would be an honor."

"Some time ago, I was given a ROK flag about the same size as a present. If we could find another pole…" said Frazee.

"Damn right. Let's get it done," Carpenter commanded.

An enemy artillery explosion hit overhead. Dust shimmered down as the men walked away.

Hill 433
July 20, 1953
0550 hours

Carpenter, Riney Unger and Da-vid were watching the radar screen. "About 150 yards out. One in front and others staggered farther back," says Da-vid. Carpenter and Riney left and joined the ROKs watching the road.

"Hold your fire," said Carpenter in Korean.

Out of the foggy early morning light came a hazy figure that resolved itself as a point man. It was an American infantry soldier. With quiet steps and with great caution, he moved forward. Carpenter called out softly, when the man heard, he crouched.

"This is Lt. Carpenter, is that you Dr. Livingston?"

"Hey! That was my line Stanley."

"Come on forward."

The man got up and came closer then saw Carpenter and Riney. He nodded and transferred his rifle to his left side and reached out to shake Carpenter's hand.

The man became aware of the odorous smell. Other Americans came forward out of the fog.

"No offense Lt., but you are rather ripe."

Carpenter and Riney laugh nervously as the both sniff their uniforms.

"Until you mentioned it, I didn't realize that. I even offend myself!"

Members of the 6th ROK Infantry Regiment approached and bypassed the position as they secured the hill. The fog quickly burned off. The remaining Americans and ROKs on Hill 433 came out of their fighting bunkers and trenches and looked around. The arriving soldiers looked clean and fresh compared to the troops that had been on the hill. Those on the hill were a dirty, smelly, ragged bunch.

A light wind helped clear some of the remaining fog, a fluttering noise was heard, and Carpenter turned. He saw the ROK flag and the American flag stirring in the breeze over Hill 433 side by side.

Trucks arrived and the wounded were brought out, some on stretchers and many walking wounded. ROK doctors in white coats were attending them. Casper helped along with the ROK medics. The wounded were loaded and left. Then the ROKs brought out the dead from below. A score of bodies were placed off to the side of the rows of ROK dead.

Occasional explosions and gunfire were heard, but most were distant. Most of the bodies had nearly pristine body bags and some were bloody with dark drying blood hiding the dead from view. Trucks arrived and the bodies were loaded up, stacked like cordwood in the backs of the trucks.

Carpenter and many of the Americans were watching the ROKs load up their dead. Carpenter went from one American body to another looking at their faces. He was followed by Whitehorse. At some, he winced from the sight. He paused at the body of Ransom. The corpse still had a slight smile on its pale face.

"They say after he got hit, he somehow kept shooting," said Whitehorse.

Carpenter nodded his head and closed the cover over the face. He went to the next body. Carpenter gave a shuddered sigh then slowly recovered the face. The last two mattress covers were distorted and tied up to keep them shut. Carpenter bent down toward them. His nose twitched. Whitehorse placed a restraining hand of his shoulders.

"You don't want to see them. Remember them the way they were."

Carpenter looked angry for a moment then nodded his head as he stood up. Without a comment, he helped load the American dead into a waiting truck.

Around the hill were hundreds of dead Chinese. Some of them were laying on top of each other 3 high. Almost all of them were facing the goal they never reached. The ROKs gathered up the very few Chinese still alive and carried them towards the bunker on stretchers.

184

ROKs with masks and gloves gathered up the dead Chinese, stripping them of military gear, and swinging them into piles every 25 yards or so.

While some piles were small with only half dozen or so bodies, many held a dozen or more. Body parts were thrown on the piles as well.

The finished "White Tiger" painting with the "blue spade" in its eye viewed the dead soldiers as they were carried by and up into the light.

With little fanfare, the FAC team of Frazee & Cushing loaded the jeep that arrived for them with their meager belongings. The jeep started and they both climbed in. It made a u-turn as Carpenter came running forward to say goodbye.

Frazee was grinning as the jeep slowed down and he threw a quarter full bottle of Jack Daniels to Carpenter. Carpenter barely caught it. Then Frazee saluted Carpenter with a wink. As Carpenter saluted back, the jeep took off down the hill.

Carpenter was still standing, watching the jeep disappear down the hill when Lt. Col. Kim walked up. Carpenter nodded at him and opened the bottle. He handed it to Kim who took a swig. Then Carpenter took a swig of the whisky.

"How do you deal with this?"

"It can be only understood by battle brothers. Those who have seen, smelled, touched and even tasted what we have done." Then, after a final swig of Whiskey,

"Just deal with it. We just find a way to deal with it."

Carpenter looked off into the distance and then back at Kim. "If I can I will, Colonel. But I will never forget."

The remains of the ROK 26th Regiment, 1st Battalion gathered into a four column formation. There were only about 100 left. Their NCO straightened the ranks. Lt. Col. Kim and Senior Sgt Park watched them. All the Americans came up to watch. They were quiet and solemn.

Carpenter stepped forward, "Where are your trucks?"

"We walked up this hill and now we will walk down this hill. The trucks will meet us at the bottom."

"Thank you for everything, Colonel."

Carpenter saluted and Kim returned it. They shook hands.

Kim then went to Sgt Whitehorse. Carpenter turned and saw that the Americans were now in two ranks with Whitehorse on the far left of the first rank and Sgt Riney in the same position of the second.

Kim went to each American who saluted then shook his hand. Sgt Park watched quietly with Carpenter. As Kim was with the last man, Carpenter looked at Sgt Park.

"Senior Sgt Park. It has been an honor." Park nodded his head then stepped back a step, coming to attention. He saluted Lt. Carpenter who returned the salute formally. Park did an about face and joined Kim who joined his unit. Sgt Cho called the ROKs to attention.

Park joined Cho with Kim facing the men. Kim said something quietly then a booming voice was heard

"Second Battalion! Forward March!"

The ROKs stepped off smartly in their columns of 4 led by their commander and their remaining two senior NCOs.

"De-tach-ment! Hand, Salute!" ordered Carpenter. All the Americans saluted the ROKs marching by. Lt. Col. Kim returned and held the salute.

"Eyes left"

"To the 26th!" said Lt. Carpenter.

"To the 26th!" said the men.

"White Tigers!" said the Koreans. Quietly, Carpenter dismissed the men

"Ready, tu. Dismissed."

The ROKs marched down the hill. They broke out in a Korean military song as they disappear down the dirt road.

"Our ride is probably lost," said Riney.

"We are still bastards. We were lost, now we are found. Now we wait," said Whitehorse.

ROKs of the 6th were still cleaning up. As the sun went down they use flares to light the gasoline drenched Chinese bodies and parts. Dozens of funeral pyres were seen. The heat generated by the gasoline ignited the clothes then the fat within the bodies as they burned. Limbs shrank and moved because of the heat. It was an eerie and gruesome sight.

The men of the detachment watched quietly and respectfully. Two deuce-and-a-half trucks grumbled their way up the hill. A jeep was leading and arrived carrying Major Hastings. Carpenter strode up toward the jeep and saluted Major Hastings as he got out. Hastings had a vivid scar on his forehead. He returned the salute. He did not recognize Carpenter or any of the men.

I'm Major Hastings. Lieutenant, where is Lt. Sharp?"

"Lt. Sharp was KIA, sir."

"Where is the detachment? I'm here to pick them up."

"Sir. What is left is here. I'm Lt. Carpenter."

As Hastings looked around at the shell holes and debris, the piles of the dead Chinese were beginning to burn.

"My God! There are only 12 of you left?"

"Yes, sir. Sgts Whitehorse & Riney, load em up."

The light wind shifted and Hastings first smelled the dirty worn men in front of him. His nose twitched. Then it flairs suddenly … he smelled the burning hair and dead being consumed by fire. Hastings gagged then threw up by the front of the jeep. He took a minute to gather himself. The detachment looked at Hastings slightly amused and slightly disgusted as they loaded up just the first truck with their meager belongings and the remains of the radars. Hastings sat down in the jeep. He was quiet, thoughtful. Carpenter stood aloof, watching.

Whitehorse looked around for a moment, then jogged off. The men climbed into the first truck. In a minute or so Whitehorse returned. He had the American flag they used. Riney reached down and helped him into the back of the truck. Carpenter turned toward Hastings.

"We are all ready, sir. May I ride with you?" Hastings nodded his head. Hastings looked like he wanted to say something but didn't. Carpenter climbed into the back of the jeep. Before he sat down, he looked around for a moment or two. Without a further word, the jeep made its U-turn followed by the trucks. The Americans left the hill the way they came. The burning Chinese funeral pyres lit up the darkening sky.

On the 17th we could see the Chinese withdrawing toward the north in large numbers. They continued to withdraw for two days. Then on the morning of the 19th, a relief column reached us. It was a Korean regiment, and a contingent of American troops from the 7th Infantry Division. They quickly moved through our position and toward the next ridge line to the north. For the first time in weeks, we were able to get out of our firing positions and trenches and enjoy the sun which was shining brightly. I had our unit pack up and get ready to leave. We had very little personal gear, nearly all of it had been lost in the pressure of our defense. We were able, however, to recover and pack away all four radar sets that we had brought to the position. We sat around all day., sunning ourselves, and watching the Koreans stack up the dead Chinese in huge piles. There were hundreds of bodies. Near the end of the day, they poured gasoline over the piles of bodies, and cremated them. That was the most gruesome thing that I have ever witnessed. I will never forget those hundreds of bodies going up in flames. Just as it was getting dark, two trucks and a jeep came to pick us up. We quickly loaded them, said goodbye to our Korean brothers, and left.

Richard Carpenter, Memoirs

Chapter Eighteen

It took all night for us to get back to the Kimpo Airport. When we arrived they immediately took our radar units away, and put us in a bunker that had about two dozen beds in it. They were all made up, and even had clean sheets. There was a shower unit, and we all got a change of clothing. I think I would still be in that shower, if I had not been so tired. They offered us a meal, but most of us skipped it and went right to bed. I did not wake up until the next morning, I had nearly slept the clock around. There was a mess hall in the bunker next to ours, and slowly as we woke up, we went there to eat. The mess sergeant served us steak and eggs, with plenty of fried potatoes, white bread, milk, and gallons of coffee. We all gorged ourselves, and promptly got sick. Our stomachs could not handle the volume of rich American food that we consumed. The mess Ssrgeant had seen this before, and really got a laugh out of it, at our expense. From then on we ate in moderation for a while.

Richard Carpenter, Memoirs

Afternoon
8th Army Command
Headquarters Seoul
South Korea
29 Aug 1953

Lieutenant Carpenter entered the General's office in an Officer's khaki uniform. He marched to exactly three feet from the desk and came to attention. He saluted the General and held it until it was returned.

"Sir, Lt. Carpenter reporting as ordered."

"Relax, son. Sit down."

Carpenter sat stiffly in the chair indicated.

"Hastings, pour us all a drink?"

"Yes, General."

The General flipped through a few pages of a folder, then closed it. Hastings handed the General a drink then gave one to Carpenter. He did not prepare one for himself. "Young man, you did one hell-of-a-job. You deserve a medal or two along with your men. BUT. We have a problem." He gestured at Hastings.

"Per the early agreements with the Commies, each side agreed to not introduce new weapon systems. Some believe we may have violated those agreements by introducing the combined ground radar mated to AA weapons. AND some fear that the Chinese would declare the ceasefire null and void."

"A renewal of hostilities is unacceptable. Your men have already been told to keep their mouths shut. They have been given the option of honorable discharges and some, ah, unofficial assistance. I assume you do not need to be ordered?"

"Yes, sir. I think I understand sir."

The General nodded his head approvingly. You got excellent marks from the ROKs. I know you want to stay in the Army. But, I have been around long enough to see the writing on the wall. The damn politicians will draw down the military. You would wind up as a second lieuie in some stateside garrison where the academy boys will get promoted over you every time. You wouldn't make your 20. That's reality, Carpenter."

"The other option is to resign as an officer. Then you can reenlist," said Hastings.

"I would like that, sir. I just want to get back to my family in Germany. I had dependent paperwork pending when I left."

"Done. Wise decision, son."

The General stood up and Carpenter quickly stood as well. As the General stepped around the desk, Carpenter placed his glass on the desk blotter and shook the offered hand of the General.

"Good Luck and Godspeed." Carpenter took a step back, came to attention and saluted. Hastings did also. The General returned the salute and Carpenter and Hastings left the office. As they walked away from the office, the Major turned to Carpenter and stopped in the hall. Carpenter stopped and faced him.

"What you did up on that hill was remarkable. I'm sorry that it cannot be told. It will be buried where such secrets belong. My advice is to forget what happened. You were never here." Hastings paused and blew some air over his bottom lip, as he stroked his chin briefly. "Again, you handled yourself very well. People did notice." Hastings paused, and glanced to see that no one else was around. Hastings used his right fore finger to barely touch the right side of the bridge of his nose, then threw the finger from the wrist at Carpenter. It is the gesture of 'just between me and you.

"You may hear from some 'Agency' friends. They pay over and above. If you know what I mean. Keep your nose clean in the meantime."

Carpenter pondered the comment for a moment. He looked at Hastings like he was going to say something. There was a bit of emotion in his expression for a moment. Then his poker face showed.

"Yes, sir. I will keep it in mind. Thank you again."

Carpenter nodded his head and offered his hand. Hastings took it warmly in both hands. Hastings did not let go for a moment or two, looking Carpenter in the eye. Hastings then nodded his head and let go. Carpenter took a step back and saluted. Hastings returned it.

"Sir."

Carpenter then turned away, walking alone down the long hallway. Hastings stood watching for a little bit, then left the other way.

Right after we ate they started to call us in for our after action reports, and debriefings. That took most of the day, and those who were waiting for, or had completed their debriefing, simply lounged around enjoying life. I remember signing my report, and it was dated July 25. We also had to sign a statement acknowledging that our operation was still secret and we were not to tell anyone what we had been doing, not even that we had been in Korea, until we were specifically released from the secrecy obligation by competent authority. We were then told that we were being returned to the States on the next available air transportation.

I guess they had a hard time deciding what to do with me. I had been commissioned. So they took me away from the other men and put me into a BOQ. I was advised that I could not wear the insignia or rank until sworn in. So I was just in limbo, neither commissioned, nor enlisted. Try explaining that to the military police. The armistice was signed, and the guns went silent on July 26th. It did not take long after that, and I got a flight out on August 1st. We refueled in Japan, and in Alaska on the way home, and landed in the States at Travis, Air Force base on August 3rd. I was then flown again to Fort Sill, where again I was debriefed.

I was anxious to return to Germany, and my family, but the limbo situation of my commission kept screwing things up. Finally, a warrant officer in the base personnel section 'came to my rescue.' If I would accept a discharge from my commission, he would arrange for me to get back to Germany. The Army did not want a high school dropout as a Lieutenant. I accepted the discharge on August 15th, having been a commissioned officer for exactly one month. On August 17th I went to Oklahoma City, and boarded a train for New Jersey. There, I was placed on a plane and arrived at Rhine Main in Frankfurt on September 1st. Then another day on the train, and I was back in Bamberg on September 2nd. That was a very happy day for me.

When I reported back to my Regiment, the Battalion Commander kept his word and promoted me to sergeant on my birthday, September 22nd.

Then he assigned me again as the leader of the M2 60mm mortar section, in the weapons platoon of "A" company. I remained in that assignment until I was ordered out of Germany as a "Homesteader" in March 1954. The Army credited all of my time from January 1948, a total of five years, as my tour of duty in Germany.

Gunda and I went through all the paperwork to get our marriage sanctioned, and in January 1954 received permission from both the German, and American authorities to again get married. This time we had a formal wedding at the City Hall, and a church wedding on post. We set the date, and got married again on February 6th to coincide with our first vows in Switzerland back in 1950. Both Laura and Linda attended both weddings.

I immediately started the court process to adopt Laura and at least got all of the paper work completed and filed with the German court before I left for the States. It was difficult for me to leave my family in Germany again, not knowing when they would be able to come to the States. But at least now we were legally married, and Gunda and the kids could use the commissary, and exchange, and American medical facilities.

My stateside assignment was to the Recruit Training Command at Fort Leonard Wood, Missouri. I took a short leave and went home to Arnegard, ND to visit mother, and then reported for duty. My assignment was as a drill sergeant in a recruit company. All the time I was at Fort Leonard Wood, I tried to get reassigned back to Germany. Laura's adoption had hit a snag. I had to have both of us appear in either a German or an American federal court to complete the Adoption. Needless to say, it was difficult to accomplish with me in Fort Leonard Wood, and Laura in Germany.

Then out of the blue, I was approached by a CIA agent, and asked to volunteer for another secret assignment in Southeast Asia. The French were getting ready to pull out of Vietnam, and we wanted to get most of the heavy equipment we had given them back. Plus, I was to learn later, we were to smuggle Vietnamese special forces teams into North Vietnam. The offer was quite lucrative, a thousand dollars a month, for a 90 day assignment, plus my military pay and allowances. But, the clincher was a promise to get me assigned back to Germany when the mission was complete.

I jumped at it, and left Fort Leonard Wood in mid-September just before my birthday in 1954. I joined a group of officers and enlisted men headed for Vietnam at Travis Air Force Base, and we were flown via Hawaii, and the Philippines to Saigon. There, I was sent to a large French cargo ship docked in the river, and shortly afterwards we moved down the river and out to sea.

On the ship below decks were two Vietnamese special forces I 10 man Teams. It turned out that part of our mission was to infiltrate these teams into North Vietnam. I learned that the teams had been trained by U.S. Forces in Formosa, and in the Philippines. We did not have much to do with them, and they stayed pretty much to themselves during the voyage.

We sailed north, a good distance out to sea, then turned toward the coast making port in Haiphong. It took us two days from Saigon to Haiphong. As soon as we docked, I was sent ashore to inspect the French vehicles and equipment that they were bringing to the dock area. It was my job to sort it out, leave the junk, and have all of the good vehicles and equipment loaded onto the ship. I had a team of about 50 Vietnamese stevedores. They would start, and run the engines of the vehicles. Tanks, halftracks, trucks, and jeeps. I would then inspect the vehicle. If it looked good, and was running, I put a large chalk "X" on it. Then the stevedores running moved it to the ship, and it was loaded. We salvaged all of the 105 mm artillery pieces we could get our hands on regardless of their condition. There was so much equipment piling up, that we did not have any trouble getting a full load for the ship.

In a fenced off area, a large number of North Vietnamese (thousands of them) were waiting for transportation to South Vietnam. As soon as we got *the ship loaded with the military equipment, we loaded about a thousand Vietnamese into the holds, and deck space. It was quite a mob of people on the ship.*

While we were loading the ship, the Vietnamese special forces teams and their equipment disappeared. Evidently they were to infiltrate, and go underground until ordered to begin operations. With the ship seriously overloaded, we set sail for Saigon. There the ship was unloaded. The military equipment was placed into storage by the MAG group, eventually to be given to the South Vietnamese.

The North Vietnamese families were simply disembarked, and walked away on their own. It did not appear that they were being given any assistance.

As quickly as we were unloaded, we set sail again, and quickly got into a trip routine. Sail, load, sail, unload. One thing that broke the monotony was the Vietnamese special forces teams. We infiltrated a team about every third trip. Another thing that helped was my Vietnamese interpreter. She was a lovely young woman, about 25 years old, and spoke English perfectly. She had lived and gone to college in the States, as a Catholic, and wanted desperately to get her whole family south. There was quite an extended family. Father, mother, aunts, uncles, brothers, sisters, and lots of cousins. About a hundred of them. When she approached me there in Haiphong, I was desperate for an Interpreter. The one that was assigned to me spoke only broken English, and had a hard time understanding my orders. I talked it over with the ship's captain, and he agreed, that I could load the family with the military equipment, before we opened the ship to the other refugees. So I accepted her services. She stayed with me throughout the whole assignment. I am sure that she was taking some kickbacks from people who were not her family, but I was so happy with the arrangement, that I looked the other way.

We made one round trip every seven days. Finally in late November

195

we started to run out of equipment to haul south, and no more special forces teams were being infiltrated. On the first of December the operation was halted, and I was immediately sent back to Fort Leonard Wood.

I purchased a car with part of the $3,000.00 bonus, and took a driving trip to North Dakota, to visit mother. That was my first Christmas at home in eight years, and I enjoyed it.

Toby had a girlfriend there in Arnegard, and she wanted to go to Minnesota. Mother also wanted to visit some relatives in Minnesota, so when my leave was over I drove back to Fort Leonard Wood through Minnesota and dropped them off.

Unfortunately Toby's girlfriend wanted to drive, and I agreed even though the roads were icy, and there was new snow on the ground. It took only a short time until and she ran off the road trying to pass a big truck. Fortunately it was a large flat area, and we only hit a rock with the right front tire. I inspected it, and it seemed OK. So I got us back on the road, and we continued the trip. We finished the trip, I dropped them off, and returned to Fort Leonard Wood.

As soon as I got back to Fort Leonard Wood, I was alerted for overseas movement to Europe. For some reason it took over a month for me to process out. So it was early February before I drove out the gate for the last time.

Toby had returned from Korea in January, and was stationed at Fort Sheridan, IL, Just north of Chicago. I drove to Fort Sheridan and visited with him for a couple of days. He was without a car, and I had no authorization to ship a car to Europe. So we made a deal, and he took over the car, and I went on by train to New Jersey, the port of debarkation. That was the first time I had seen Toby since I left North Dakota back in 1947.

I was quickly put on a roster, and then a plane for Europe. I was *very excited. The assignment I was given was to the 4th Infantry Division which had their headquarters right in Frankfurt.. So when I got off the plane, a group of us was bussed directly to their headquarters, processed, and given assignments. My Assignment was to "E" Company of the 8th Infantry Regiment. This was my second assignment to "E" Company. It was in "E" Company that I took my basic training back at Fort Ord, California. The next day, I was bussed to the company which was stationed in a small town called Budigen, only a short distance from Frankfurt. My assignment was as platoon sergeant for the weapons platoon. The Platoon had two sections. One of 60 mm mortars, and one of 57 mm recoilless rifles.*

The best part of the assignment however was that new government quarters were available, and I had sufficient rank to be assigned a unit. The company commander understood my situation, and gave me a week's leave, and the use of a jeep and trailer, to move our household goods. I set off immediately for Bamberg. Had a fine reunion, packed everything up, and returned to the quarters in Budigen. We used most of the week to get settled in, and to get the girls enrolled in school. Then it was off to work.

The weapons platoon was mostly made up of a buddy platoon. A group of twenty men who had enlisted together, and had been promised that they would stay together while completing their service. I have never before, or since found such a cohesive, cooperative group of men.

Whenever we got into a competition, regardless of what it was, they put everything into it, and we usually won. Every time one or more of them went on guard duty, one of them became the Colonel's orderly. Every weapons competition, weapons firing, Inspection, etc. we always did well, and accumulated a significant number of trophies, and accolades. This included during our cold weather training at the Wildflecken training area, high up in the mountains near Fulda.

This was one of the better assignments that I had in the service. An excellent job, good troops to lead, a close by family, and an ability to walk to work. Two things about the family stand out in my mind about Budigen.

One was that one of my men found a fawn when we were out on a field problem, and brought it back to camp with him. Unfortunately, after it had been handled we could not release it back where it had been found. The mother would have rejected it. So I took it home, and it became a household pet. We had a cage enclosure for storage in the basement of our building. That became the deer's home. We named him Willie, and all of us became quite attached to him. When we left Budigen he was getting to be pretty big, nearly a yearling, and we could not take him with us. So I made arrangements with a German Forest Jager to take him to a small park zoo. We all felt bad about leaving him behind.

The second that stands out was a visit from Gunda's mother. One day she walked down into the town to do some shopping. It was only a short time until she was back, and appeared to be upset. We asked her what had happened, but she would not tell us. Later she confided in Gunda, that: "None of those Americans in town could speak German!" We all got a good laugh out of that, as all of the 'Americans' in town were actually Germans. She had simply run up against a different German dialect. In early 1956 the 4th Infantry Division was to rotate back to the States, and be replaced by the 3rd Armored Division. A lot of us who had not completed our three year tours were to be reassigned. My reassignment was to the 11th Airborne Division, and the 370th Armored Infantry Battalion which was attached to them from the 7th Army. In May I was ordered to Munich, where the unit was stationed, and where quarters were available for us.

I reported in and received my assignment as the Armorer for "C" Company, and responsible for all of their weapons. Plus I also was in charge of all their ammunition. It was quite a job. After explaining my situation to the company commander, I was given leave, and a truck to move my family to Munich.

Gunda was pregnant again, with a due date of June 15. We wanted to get her to Munich before she delivered. Unfortunately, it did not work out. As I was approaching Budigen on the road, she was in an Ambulance headed to the Hospital in Frankfurt with labor pains, and we passed one another without knowing it. Fortunately Gunda's mother was staying with us, and had remained with the children. As quickly as I got to Budigen, I was advised of the situation, and turned right around and went to Frankfurt. I arrived there just in time to welcome our son John into the world. It was Memorial Day, May 30th. We named him John Robert, after my stepfather John, and my brother Robert.

Gunda stayed in the hospital for several days. We used the time to pack up, and clear quarters. So that when she was released, we simply went to the Bahnhof and took the train to Munich. I had sent our household goods on in the truck, and Gunda's Mother accompanied us to watch over Gunda and John.

(NOTE: The computer crashed again and the hard copy was lost for many years. My son John OCRed the report and set me back to work on this again!)

I have now gone on to other things, and have a hard time getting back to continue this biography. One day I will get on with the "Rest of the Story."

Epilogue

The Top Secret Battlefield radar would not become
public until 1959.

The story of the brutal combat action on Hill 433
remains one of many untold stories of the Korean War. Until Now.

A final peace treaty has never been signed and
technically the two Koreas are still at war. No medals or
recognition was given for the valor of the men involved at
the heroic stand on Hill 433.

I traveled to meet the Carpenter family in San Diego in
September of this year. On a pleasant, warm, California afternoon, we
sat around a backyard patio and got to know each other. After a few
glasses of wine, everyone loosened up and the stories of Dick Carpenter
began to flow. I met all of the children, that were still alive, and got to
know them.

I especially enjoyed meeting the girls, Siggie and Ruth, who took
the time to pull me aside and tell me their experience of their father.
Siggie seemed especially emotional about the gathering. She told me of
how her entire life, she never felt like she belonged. She looked different
than the other kids. She recounted how the German side of the family
told her that Dick was not her real father. When she confronted her
parents, they denied it. Dick told her, "You're my daughter and that's
final."

Siggie told me how this feeling of not belonging was with

her during her entire life and had caused problems in her personal relationships. After Dick died, the children had a DNA test performed and sure enough, she was not his biological child. This caused her even more pain, rather than closure, as it left many questions unanswered. "Why didn't he just tell me?" she asked me out loud.

Later in the evening, as the get-together wound down, the children were all seated around the outdoor table. The sun had set but the air was still warm, as it seems to always be in California. It was at that point that John, the eldest, told Siggie that he had found information that her mother was indeed raped during one of Dick's extended absences for duty. "They must have made a pact to never tell any of us," John said. "He would have never broken that promise to her."

Tears rolled down Siggie's face. "Thank you, John. That is the closure I have waited for my whole life." The pain that had been written on her face seemed to wash away from her.

I turned to Siggie and said, "Siggie, after all you went through, the not knowing, it seems you were a lost bastard as well."

"Yes," she said with confidence. "I guess I was."

My father formally retired after 41 years of combined military (as a Major USAR) and other service to his Country in 1987. Gunda, his dear wife died in 1990. In 1992, he retired from the County of San Diego. Most people that he worked with in the County had no clue of his side-work. And that was the way it should have been.

Richard L. "Dick" Carpenter died in his sleep from heart failure on 4 January 2013 in San Diego, CA.

John R. Carpenter

About the Author

L. Todd Wood is a graduate of the U.S. Air Force Academy. He has been an aeronautical engineer and an Air Force helicopter pilot. In the Air Force, he flew for the 20th Special Operations Squadron, which started Desert Storm. He was also active in classified counterterrorism missions globally supporting SEAL Team 6 and Delta Force. For eighteen years, he was an international bond trader with expertise in Emerging Markets. He has conducted business in over forty countries. Todd has a keen understanding of politics and international finance. He is a national security columnist for The Washington Times, and has contributed to Fox Business News, The Moscow Times, NY Post, Newsmax TV, Breitbart, Zero Hedge, and others.

Other Works by L Todd Wood

CURRENCY

SUGAR

DELTA

MOTHERLAND

Dreams of the Negev, A Short Story

The Last Train, A Shorty Story

Please post a review on Amazon if you enjoyed the book!

Find out about new releases by L. Todd Wood by signing up
for his monthly newsletter found on LToddWood.com

Connect with Me Online

Twitter @LToddWood

LinkedIn (L Todd Wood)

Facebook (L Todd Wood, Novelist, Correspondent, Pilot, Bond Trader)

Pinterest (L Todd Wood Author Page)

Google+ (L Todd Wood Author)

LToddWood.com

LostBastardsBook.com

CURRENCY

by

L. Todd Wood

Prologue

Weehawken, New Jersey
July 11, 1804

The smartly dressed older man came first, sitting erect and still as death in the rear of the long oar boat as it silently rowed across the wide river. The moon cast an eerie glow across the fast-moving, silky, black current.

He was balding, middle-aged and had dark features. However, he was in a much darker mood, a murderous mood in fact. He was the kind of man that never forgot anything; especially, the stain on his honor. His eyes bored holes in the back of the man sitting in front of him and he did not notice his surroundings as his mind was lost in thought. He was there to right a wrong he had suffered.

To this end he was joined by two other men seated near him, as well as two additional young rowers and his dueling second at the head of the craft, a total of five. The only sound was the water lapping like a running brook as the oars slipped in and out of the calm silvery surface. Slowly the boat crossed the dark current. Preoccupied, the passenger did not hear. He was focused only on the task ahead of him.

They beached the long oar boat upon the bank and he and the three men quickly scurried into the woods as the rowers stayed behind. Immediately the four gentlemen began to clear the brush along the ledge facing the water. The birds awoke but no one heard. Their singing cast an odd joyful sound, contrasting eerily with the morbid events unfolding beneath them.

A man younger by a year arrived a half hour later in a similar craft with a smaller entourage. He was a person of importance and seemed rather arrogant. In fact, he had a brilliant mind. Unfortunately he had a habit of taunting others with his brilliance that brought him to where he was at this hour. His pompous mood seemed out of touch with the somber circumstances.

One of his party was a well-respected physician. His second, sitting in the bow, carried an ornate box the size of a breadbasket. Inside were two Wogdon dueling pistols, the finest in the world at the time. The pair of weapons had already claimed the lives of a handful of men. One of these killed had been the younger man's son.

The first party made themselves known and the group just arrived made their way up the embankment to join them. Salutations were exchanged.

The seconds set marks on the ground for the two men ten paces from each other. The younger man, since challenged, had the option of choosing his spot and had already selected to be facing the river. The two antagonists loaded their pistols in front of the witnesses, which was the custom, and the seconds walked into the woods and turned their backs. This way they would not be party to the scene and could not be charged with a crime as dueling was now illegal. The honorable gentleman was becoming a rare breed. Times were changing.

The blonde man's second began counting down. Unknown to his charge's opponent, the pistols had a secret hair trigger firing mechanism; just a slight application of pressure would ignite the powder. This was a slight of hand to say the least.

A loud crack rang out. A few seconds later, another. Then a cry of pain. Whether the younger man accidentally fired due to the hair trigger or intentionally wasted his shot, we will never know. Historians have debated this point ever since. His shot missed his adversary and ricocheted into the surrounding trees.

The return fire from his opponent however was deadly. The ball pierced his abdomen and did mortal damage to his internal organs before lodging in his spine. He collapsed to the ground.

The acrid smell of gunpowder still hung in the air as the dark haired man walked up to him writhing on the ground. He was confident in his errand as he stood over him and methodically reloaded his pistol.

"Where is it?" he asked as he calmly packed the powder down the barrel.

The seconds stepped forward out of the brush but the older man waived them off with his pistol. The New Jersey woods were strangely quiet; the New York lights across the river twinkled in the background, soon to be obscured by the rising sun. Its rays would soon shine a bright light on the deadly events happening below.

"Where is it?" he said again sternly but softly, pointing his reloaded pistol at the man's head as he tried to lift it off the ground and speak. The long highly polished brass barrel reflected the early morning sun.

Blood poured from an open wound in the gut. Although mortally wounded and lying in the dirt, he held his hand over the opening to try and stop the flow.

"Go to Hell!" he gurgled as his mouth filled with blood.

"I probably will but I think you will beat me there," the darker gentleman chuckled and knelt down beside him. He started going through his bleeding man's pockets. "I have heard you always carry it with you." Aaron Burr knew he didn't have much time before the surgeon and seconds gathered and pulled him off. Inside the man's blood soaked gathered and pulled him off. Inside the man's blood soaked coat, he found it.

"Ahh!" he gloated smugly. He quickly hid the pouch inside own vest and stood.

"You will never find what you are looking for!" the wounded gentleman said in a whispering laugh. His strength was ebbing. He was going to die.

"We'll see," replied Burr.

"He's all yours!" he called to the second and the wounded man's supporter rushed forward and tended to Alexander Hamilton.

Chapter One

June 10, 2017

Bahamas

The seawater thundered over the stern as the old fishing boat attempted to cut through the eight-foot waves, soaking everyone on board to their core. Connor Murray felt the impact in his kidneys as he held on to the ladder for dear life. The tuna tower swayed above him. His arm muscle burned as he prevented himself being tossed into the sea from the violent movement. Saltwater stung his nose. The new motor growled like a wounded bear as it strained against the onslaught.

How long can this go on? he thought.

Connor had puked twice and didn't relish a third attempt, but the nausea in his gut and throbbing in his head told him it was inevitable. The other two weekend warrior fishermen with him on the stern were leaning over the side as he contemplated his situation. He heard them groan as they tried to empty their stomachs, but there was nothing left inside them. He was miserable.

The day had started easy enough in the predawn hours as they boarded the boat at the end of New Providence Island east of Nassau. The water was calm this early in the morning. This part of the island was protected by natural reefs. The rapid change in depth as the ocean floor rose to the island caused the ocean and the island currents to crash into each other and protected the east end from the ocean's wrath.

He had been planning this trip for weeks with his good friend. Alex, his Bahamian business colleague, had just finished overhauling an old, thirty-two-foot pilothouse cruiser, a labor of love for two years. The boat had been refurbished from bow to stern; many a weekend night they had spent drinking beer on the deck after a day's work on the "yacht, "as Alex's wife called it. Conner chuckled to himself. She was a little bit of a "wannabe."

The boat was not pretty, but she was strong. Alex had seen to that. He took very good care of her. When someone tears down and rebuilds something that intricate, it becomes part of them. The boat had become his passion. His wife didn't seem to mind; she had her yacht.

Connor had met his friend years ago when Alex was the head trader at a large hedge fund located in the tony Lyford Cay area on the west end of the island.   Alex mingled with the movie stars. After trading together for several years and socializing on every trip Connor took to Nassau, they became friends and trusted one another completely. They took care of one another as they both moved firms several times over the years. Their careers flourished and their wealth grew.

This trip was a much needed change of scenery away from the Bloomberg terminal for both of them. The constant movement of the ocean was a welcome relief from the volatility of the markets. Stress relief was critical in their business. A day or two in different surroundings did wonders for one's trading acumen. "The three-day weekend was invented for Wall Street," Alex would often say.

The sun was rising as they cruised past old Fort Montague close to East Point. He could make out the row of old, British twenty-four-pound cannons lining the top barricade. The fort had held up remarkably well through the centuries, considering it was constantly exposed to the elements. The history here ran deep. Connor loved it. The islands were in his blood now. The open ocean waited for them ahead.

Today the only possible negative was the sea, as there was not a cloud in the sky. Their luck was not good.

The trip out of the channel was hellish as they crossed the churning waves. The barrier reefs produced a literal wall of water the boat had to climb, as the underwater structures halted the ocean's momentum. It was no better on the other side, but Connor knew Alex would not turn back. He had promised everyone on board some fish and this was her maiden voyage. They all quickly decided it was too rough to make the two-hour trek to Exuma, which they had planned. Instead they would stay off Paradise Island. Connor was quietly grateful.

Later in the day, after they had landed three dolphin⊠a bull and two females⊠all mercifully decided that it was too rough to continue. Now they were making their way back to protected waters, but it wasn't easy. The waves beat the side of the boat and refused to provide a respite to the passengers and crew. The fishermen were still hugging the rail in case they had to empty their stomachs again.

Thinking he would feel better higher, Connor climbed up the ladder and leaned into the back support next to Alex. He could see a line of boats making their way into the harbor to escape the violent sea.

The tower swayed violently as Alex strained to control the craft as she muddled her way up and over the crest of the waves.

"Must have taken on some water," Alex growled. "I can feel it sloshing back and forth down below. It's hard to control this pig."

Connor wasn't listening. "Tell me the latest on our discovery," he said.

April 23, 1696

Captain William Kidd turned one last time and looked over his shoulder at the Plymouth landmass disappearing in the background. The city that produced the Pilgrims had long since lost its dominance as a critical port for the Crown. The coastal lights faded in the mist.

Kidd would not miss England. He would, however, miss his beautiful wife of five years, Sarah, and their daughter in the English colony of New York, recently acquired from the Dutch, which called her New Amsterdam. The city was on its way to becoming a cultural and economic center of North America.

Sarah was one of the wealthiest women in the New World, primarily due to her inheritance from her first husband. She had been already twice widowed and was only in her early twenties.

He had left them several months ago to fulfill his dreams. He would not see them again for three years. Such was the life of a sailor.

This voyage has started very badly, he thought to himself. Maybe it was a sailor's intuition, but he had an uneasy feeling in the pit of his stomach.

He turned his attention back to the Adventure Galley. He had overseen her construction himself in London. She was built in record time, so there would be leaks and other imperfections to deal with but she would do the job. Kidd had sold his former ship, Antigua, in order to raise funds. The King was lusting for pirate blood and treasure. He made no secret of his haste for Kidd's voyage. The new ship would have to do.

She was a strong, thirty-four-gun privateer and would be formidable when he engaged pirates. Her design was elegant at 284 tons. Her oars would be an important capability when maneuvering against an enemy. His crew, however, was another story.

He had tried to leave England several weeks earlier with an altogether different group of men. His mission was financially speculative. The men would be paid the prize booty they could seize from legitimate pirate or French ships. Therefore he wanted men of good character who were excellent seaman. Having personally chosen each of the 150 men, he took pride in his selections.

In a hurry to depart London, he had chosen not to salute several Royal Navy vessels leaving the mouth of the Thames. It had been a mistake. This was a long-standing tradition and a direct affront to the English Navy and the King. His men had even taunted the English yachts as they passed, showing their backsides. Hence the Adventure Galley was boarded, and thirty-five of her best seamen were pressed into the Royal Navy. It was another several weeks before he could get Admiral Russell to return sailors to him to fill his crew. He received back landsmen and troublemakers rather than the original able-bodied seamen.

He was now on his way to New York to fill out his crew with another eighty good men. Then he could chase pirates.

"My crew will not like to be sailing under an unlucky captain," he said aloud. Perhaps it was nothing to be worried about. "Maybe we have gotten the bad luck out of the way at the start."

He turned around and faced the bow.

The open sea helped calm his nerves.

Captain Kidd was very glad to be leaving England in command of his own powerful ship. The crew could be dealt with over time. He was a restless man.

He had desperately wanted a commission from the King to command a Royal Navy vessel and had sailed to London from New York in late 1695 in search of this honor. He loved the sea and had been a respectable member of New York society for several years. He had used his considerable maritime skills as a merchant seaman to build his wealth. Kidd was in love with his beautiful Sarah and their young daughter of the same name, whom he adored. But, his first love was the sea. He wanted adventure.

He had sailed to London with a recommendation letter to request an audience with the King in his quest to become a Royal Navy Captain. While there, he became involved in a scheme to help the monarch with his pirate problem while making money for himself and others. Several financial benefactors backed him in building a ship and outfitting it to sail against any pirate he could find, with the booty paying the mission's expenses. The profits would be split among Kidd and the powerful English gentlemen. These included the Earl of Bellomont and other aristocrats. He never met the King, but he did receive a written commission to perform this duty. The King was to receive ten percent of the take.

This was a dangerous gamble, and Kidd knew he was sailing in treacherous territory. He was already at odds with the Royal Navy, who presumed it was their duty to deal with the piracy issue. It was an unlucky start indeed.

He had no friends at sea, and he suspected as much as he crossed into the Atlantic. The rewards, however, could be great and in his mind worth the risk.

Looking out over the vast ocean, he felt at peace for the first time in many years. His wife and daughter were the furthest subjects from his mind. He was with his love at last.

Chapter Two

March 8, 1806

The West Virginia snow was falling heavily and had accumulated a blanket over the lawn of the mansion two feet deep. The trees sprang from the soft surface and reached to the snow clouds above with white-covered arms. The silence was deafening in the late evening. Aaron Burr stood in the study, looking out over the magnificent gardens, which were now covered in a winter coat. The moonlight reflected off the new fallen snow. It seemed as though they were completely alone in the Virginia woods, but he knew the staff and guards were posted elsewhere. The clock struck eleven.

He had just finished a wonderful meal with his hosts, the Blennerhassets, in their formal dining room in the main section of the sprawling home, the grandest structure in all of America away from the East Coast. The recent stroll under the covered walkway to the study was short but exceedingly cold. He was thankful the servants had started the fireplace hours ago, so the room was comfortably warm. The only sounds were the crackling of the embers behind him. He listened to the popping and hissing and was comforted by long-forgotten memories of the same. He momentarily was brought back to his childhood, being cuddled by his mother next to the fire. It was a pleasant memory.

The owner of the study was obviously an educated man. Musical instruments occupied the corners of the room. The walls were covered with books from floor to ceiling. What was most interesting, however, was a complete chemical laboratory opposite where he was standing, one of two in the entire American West. Blennerhasset was a man of many talents.

Burr was excited. His plan, which until recently had been just that, a plan, was now a distinct possibility.

After Hamilton's death, he was indicted for murder in New York and New Jersey. Although he was eventually acquitted, his career as a politician was over. Yet he still craved two things, money and power.

The only place he surmised he could acquire both was out west. When he actually thought up this grand design was unknown even to him; it had been percolating in his mind for as long as he could remember. He was a very ambitious man. He had been vice president of the United States for goodness sake. Close but yet so far.

Burr had been spurned many times before. He harbored grievances. It started with General Washington, who refused to acknowledge his bravery in the Revolutionary War. Then other dreams had been taken from him.

So his thoughts had turned west.

The Spanish lands in North America were very poorly managed; everyone knew that. What he desired was no less than conquering these lands during the upcoming war between the United States and Spain. Then he would install himself as king.

He had leased forty-thousand acres from the Spanish government in the Bastrop lands of Texas along the Ouachita River in what is now

Louisiana. There he had a force of eighty men encamped, the start of an army.

I will rule benevolently, he thought.

To this end he had been contacting prominent people who he thought could help with his quest. Harman Blennerhassett was one of those individuals. Burr had appealed to his vanity and his greed. It had worked.

He was a wealthy immigrant from Ireland who controlled a large island in the Ohio River. His home was the most magnificent structure Burr had ever set foot inside. The seven-thousand-square-foot mansion contained oil paintings from Europe, silver door hardware, exquisite oriental rugs, and alabaster chandeliers with silver chains. His estate consisted of the entire landmass of the large island surrounded by the river.

Burr surmised his host wanted more money and a little glory. He was right. Tonight's dinner had sealed the bond between them and access to the resources Blennerhassett could offer.

Of course, Burr had brought his daughter Theodosia to Ohio with him. She was an asset when it came to impressing moneyed interests.

He had raised her as a prodigy with strict mental discipline. Fluent in Latin, Greek, and several other languages, she could intelligently converse with anyone. She was also skilled in the arts of dancing and music. Burr had seen to her education personally since his wife had died years ago. He loved "Theo" desperately.

She was also married to Governor Alston of South Carolina. After the birth of their son in 1802, her health became frail. Burr hoped this trip would help restore her strength. After dinner this evening, she had retired early, which allowed him to speak freely to Blennerhassett.

His host's money would help, of course, in the early days of his scheme; Burr had no resources of his own to tap. The island grounds would also be a convenient training ground for the invasion force Burr was planning. In the long run, he would need serious money, money to fund an empire.

Burr smiled. Thanks to Alexander Hamilton, I will have this. A noise startled him and he was brought back to reality. Someone was walking through the covered walkway from the main house to the study. He heard the door open and saw his host walk in with a smile on his face.

"You have quite excited my wife, Colonel. It is a feat even I have not been able to accomplish in quite some time. My congratulations!"

"It is good to have such believers in my capabilities, Harman. Let me congratulate you again on these splendid surroundings."

"And your daughter, Sir, she is an exquisite creature!"

"Thank you, my new friend," Burr replied.

Blennerhassett was uncorking a very nice bottle of Bordeaux his servant had retrieved from the cellar beneath the study.

"I drink to the surroundings to come! I must endeavor to learn Spanish," he quipped.

"Indeed," replied Burr as he raised his glass.

June 10, 2017

Bahamas

Evening

Connor parked the rental car late in the evening and walked the short distance to the British Colonial Hilton in Nassau, where he typically stayed. Nassau was very hot this summer night, and he worked up a sweat as he reached the hotel. The traffic was still heavy on Bay Street as the tourist revelers hit the bars and mingled with the locals. The scene reminded him of wolves attacking the shepherd's sheep. Many a pocket was picked as dusk descended on the pirate town.

The Hilton was built on the site of an original British fort torn down at the end of the nineteenth century. A hotel designed by Henry Flagler, who built the Breakers in Palm Beach, had replaced another military structure that burned in the 1920s. The Bahamian government then rebuilt the site, and it was taken over by Hilton in the 1990s. Hilton had preserved the colonial essence that history called for. It was a great base of operations for Connor, as it sat in the middle of his clients in Nassau, and he enjoyed working from the hotel.

The monstrous hulks of three cruise ships parked on the sea side of the building dominated the harbor and lit up the evening sky. The government had just dredged the channel to accommodate this new line of massive vessels. Their stacked decks illuminated the entire Bay Street area like a glowing sun. Even with a bad economy, the pleasure cruise business was humming.

Connor had long grown used to the ships in the background, and the sight no longer startled him. He was tired from the day's trip as he arrived at the hotel, but his mind was racing.

The doorman opened the glass door, and the welcome blast of the air conditioning hit him full force in the face. The great mural of the town's past dominated the far wall above the massive stairway to the second floor.

The history of the Bahamas and the Caribbean always fascinated Connor. Although the Bahamas was not technically the Caribbean, he lumped them together all the same.

Most Americans had no idea of the battles that had been fought here for control of the world by the colonial powers right off their doorstep. Slavery, sugar, silver, and gold were all reasons the Spanish, French, Dutch, and English came to blows over several centuries in the West Indies and Caribbean Sea. Although Nassau was never invaded, there were many forts built to protect the port from foreign powers or, more often than not, pirates.

With the immense gold and silver harvests the Spanish mined from the Spanish Main came intense interest from competing powers and other groups intent on relieving them of their precious metal burden as it was transported back to the European continent. The Caribbean islands and the Bahamas, with its vast network of isolated cays, was well suited for pirates and privateers who were determined to pillage Spanish assets as they sailed north to Madrid.

When they were not attacking foreign ships, pirates would set up encampments on the cays and gorge themselves on their newly acquired plunder. These dry spells could last for months, until another hapless victim sailed by. They loved to roast the meat of local wild pigs in thin strips to survive. This food was called "bouchon" after the French word for roasted meat. This was the origin of today's bacon. Also, since they subsisted on this ration, the pirates earned the name "buccaneers."

Locations in Jamaica and Cuba specifically became famous for their pirate cities where the thieves could congregate between exploits and spend their ill-begotten wealth. Port Royal and Tortuga were notorious examples. Nassau was also one of those places. It was one of the last pirate refuges before the European navies were able to all but snuff out the pirate vocation in the early eighteenth century.

Men such as Edward Teach, or Blackbeard, and Captain Henry Morgan became famous during this golden age of piracy. Teach was killed by English troops, while Morgan famously died in his bed of old age and a very rich man. He considered himself a privateer, but his actions drifted into the realm of piracy over time.

As the Spanish empire waned and their mining operations slowed, the region became famous for another commodity—sugar. The agreeable climate along with a plethora of imported African slaves made the Caribbean the perfect area to grow the cane to sweeten the cups of the numerous coffee houses across Europe. The indigenous Indian population had been decimated by disease upon arrival of the white man and could no longer be counted on in sufficient numbers by the slave masters.

The difference between a pirate and a privateer was slight. Pirating was illegal worldwide and a scourge on trade across the globe. However, a privateer was basically a pirate sanctioned by a government. If, say, England was at war with France, the two governments would sanction and fund captains to capture and pillage ships of the opposing power. The end result was the same. People were killed or imprisoned and ships and cargo plundered.

Connor walked across the expansive, marbled lobby and strolled into the bar facing the harbor. He had showered and changed at Alex's place after the miserable fishing expedition. He now just wanted a drink and a quiet place to think. A dark, empty corner worked just fine.

Alex had informed him of some exciting developments, and now they had to plan the next steps, which would not be easy, considering the typical government interference with situations like this one. The one thing he was quite sure of was that he didn't place any faith in the Bahamian government's fairness or process of confidentiality.

He pulled a notebook from his briefcase and began to make plans for the next few days. He tried to keep his mind busy, but again, as always, the sorrow began to creep in. It was always in late night bars that it tended to happen, when the day was winding down. That's when he thought of her and how much he missed her. They had sat in this very bar together. He shockingly realized that probably on this very sofa he had held her. She had traveled with him frequently in the past as he visited clients. The Bahamas had been a favored destination for both of them. He tried to force the thought away and concentrate on his work.

She had died on 9/11. He still had the message on his phone. He played it from time to time.

Connor ran an emerging market trading desk for a large investment bank and was based out of New York City. Although he oversaw many traders and salesmen, his main job was building solid relationships with clients, so he traveled a great deal. Many of these clients were based in the Caribbean, where he had focused his career for the last twenty years. On September 11, 2001, he was in Jamaica. He had returned to his hotel later that evening after a fishing trip with a group of bankers, oblivious to the carnage being perpetrated on New York. He had immediately tried to call her, but there was no answer. Then he saw a message on his mobile phone.

"Connor, something's happening! There is smoke all over the place. I'm scared but I love you so much!"

He never saw her or spoke to his wife again. Much later, he spoke to the police about what had happened to her. It seems most of the people on that floor were unable to get down the stairs blocked by wreckage. She probably jumped, as the fire was intense. He shivered again. The emotional pain was still raw.

He realized again that a tear was rolling down his face. "It's been a long time," he said aloud. "You need to move on with your life. She's gone," he added.

Signaling to the waiter, he downed his drink and ordered another. It's going to be another long night. There were several women across the bar who kept stealing glances his way, but he wasn't interested, had not been in a long time. Work was his only pleasure these days, but it was starting to get old.

The second drink helped, and he realized he could not get much done late in the evening. He would need his strength for the following day's activities, so he folded his notebook and picked up the letter from his great aunt he had received from an attorney in Nassau a month ago. He vaguely remembered her face, as he had not seen her for probably forty years. She died when he was a child. Even then he mainly remembered her special cakes with the icing on them that he loved so much, and the large kitchen with such wonderfully strange utensils in that old, Victorian house. He was quite shocked when the attorney for the trust had called.

"Your Aunt Clara selected you as trustee for an offshore trust before she died. She didn't want this letter delivered until you were forty-five, and it's yours now. Please come to Nassau, and let me hand deliver it to you as she requested."

"Dear Connor," the letter read, "I want to tell you about your relationship to Aaron Burr."

DELTA

By

L Todd Wood

Prologue

She was thirty-eight years old and a virgin. Her parents had seen to that. They had selected her when she was only ten to be the guardian of the flame. Her life was laid out in front of her before it even started. It had not been a bad life; in fact, it was quite pleasurable. She was worshipped and held a very high position in Roman society. She even had her own box at the Coliseum with the five other virgins. But, as with the clouds in the sky, things always change.

Julia and the five Vestal Virgins guarded the flame in the Temple of Vesta. The virgins and their ancestors had guarded the flame for a thousand years. The temple was a fifteen-meter-wide, circular edifice in the Foro Romano, supported by twenty Corinthian columns. It was one of the oldest structures in Rome and was used to store important records and business documents for safekeeping. There was an opening to the east pointing towards the sun, the origin of fire. The flame burned continuously inside. It was said that if the flame ever went out, Rome would fall. The year was 298 AD. Vesta was the goddess of fire, the goddess of the hearth—the fire that kept an ancient home alive. She was worshipped originally in the circular huts the Roman tribes built in the area, hence the circular design of the temple. The goddess kept Rome alive as long as they kept their covenant with her to keep the flame burning. At least, that was what the people were led to believe.

Julia also had a covenant with Rome, although not of her choosing. Her parents had offered her as a virgin to guard the flame when she was ten. The virgins came from very high-placed families in Roman society. It was an honor to have a daughter selected to guard the Temple of Vesta. In return for thirty years of celibacy, upon their fortieth birthday, the virgins were allowed to marry and received a huge dowry from the state. They had statues made in their likeness that were placed in the gardens around the temple. However, if a virgin broke her vow of celibacy to the Empire, the consequences were dire.

The Vestal Virgins lived in a multi-room structure right outside of the temple. The site was the most holy in Roman culture and was placed squarely in the center of Foro Romano, where it all began. This was where the first tribes of the ancient valley met to trade along the lowlands of the river. It was where Romulus was suckled by the she-wolf after being abandoned by his parents. Any free Roman citizen could take the fire to his home, and the temple therefore represented the hearth of Rome.

It was early evening when the visitor came to call on Julia. He was a younger man, a servant dressed in servant's clothes, and quite handsome. She met him at the gate to the temple grounds to talk after he had sent a request in to the College of the Vestals to speak with her. The senator's aide could come no further. "Tonight at midnight, Senator Thor will pay you a visit. He has something to give you, something that needs to be guarded, even from the emperor himself. This is the safest place in Rome. Please meet him." The visitor left without explaining further. Julia was left wondering at the gate for some time but finally retired to her room.

Julia was troubled. She would have to be very careful. This meeting was very dangerous for both her and the senator. She knew things were changing in Rome. The corruption was rampant. The emperor was claiming for himself more and more power. The Roman order and process that had survived for centuries was giving way to raw corruption and tyranny.

The Senate had long been relegated to the periphery. Originally the body was set up by the early Roman kings and came from the historical group of elders the tribes organized to help govern themselves. In fact, the word senate is derived from the Latin word senex, which means old man. Once Rome became a republic, the power of the Senate grew exponentially. However the republic was long gone. All power was now held by the emperor. No longer was he seen as an equal to the average citizen in Rome; he was a god. However, he was becoming more and more corrupt, cut off from communication with his subjects and events throughout the Empire. He received information filtered by his court with which he constantly feared revolt and death. His actions were not those of one concerned about the future of the Empire but of one concerned with staying in power. While he concentrated on giving out favors, the barbarians advanced to the north.

Night fell. At the appropriate time, Julia rose from her bed and left her chambers, moving as quietly as possible. She made her way out into the warm night. She could see the light from the flames of the hearth in the Temple of Vesta licking the ceiling of the ancient structure. She was scared. However, she trusted the senator and knew he was a good man; she would meet him despite the danger.

She made her way silently across the garden between the wading pools and stopped near the stone fence on the other side. Her white evening clothes stood out like a ghost under the full moon. The cicadas sang a rhythmic song of joy to the white orb in the sky. The Roman Forum was silent.

"Julia," a voice whispered. "I am here." She turned and walked toward the sound. The senator stepped from the shadows. "Thank you for coming," he said softly. He was old, probably over seventy, which was ancient for a Roman man. His eyes were flanked by deep crevices in his skin, and his hair was a wispy white. He walked with a pronounced stoop. He was dressed in a tunic made of expensive cloth with large, colorful stripes, identifying him as a senator.

"I came as you requested. What is so important? I cannot stay long," Julia declared.

"I don't have much time," he said and handed her a small, stone tube typically used to store documents. It was capped at both ends and sealed. "You must guard this with your life. It is the past and the future of Rome. Do not place it in the temple with all of the other royal documents. Keep it with you at all times. Tell no one. It is safe here I believe. No one will bother you. When you are older, pass it on to one of the other virgins with a sacred oath to guard it with her life as I request you to do."

She took the container. It was surprisingly light. There was a lanyard attached at both ends. She put the cord around her neck, and the scroll dangled between her breasts. She moved it under her night clothes so it could not be seen. "I will do as you ask," she replied, "because I believe you are a good man that wants what's best for Rome. I have seen you fight to restore Rome to its former glory and justice. I trust you." Julia had heard of the senator's reputation as being kind and wise, although they had never met in person. She looked him in the eyes one last time then looked around the courtyard, frightened that she would be discovered. "I must go."

With that reply she turned and walked back across the gardens. The senator disappeared into the night. What neither of them saw was another young girl, barely fourteen, also in a white night dress in the garden, hiding behind a column at the Temple of Vesta. She was Cornelia, one of the other five virgins. She had been guarding the flame through the night. Cornelia was still young but quite adamant in her opinion of herself and the importance of her role in Roman society. She was new to the temple and rather passionate about her duties. Cornelia was also a budding woman and felt subconsciously angry about not being able to talk to the young boys she frequently saw looking at her. She was jealous of the other girls in society and really quite angry about it. She knew that it was not permitted to meet male Romans without a chaperon, especially at night, and decided to take her anger out on Julia, as it was the only avenue she had, however misguided. Had Julia broken her vow of celibacy to the temple? She would need to report this to the authorities. She was certain she would be well rewarded by the emperor.

The next day, Julia rose late in the morning after a fitful sleep. She had not slept well after the meeting with the senator, and sleeping with the scroll draped around her neck would take some getting used to. She moved to the window and drew the curtain to let in some light. Strange, her morning servant was not next to her when she woke. That had never happened before. Her servant had taken care of her every need flawlessly for years. Since she had slept late, she was hungry. There was no breakfast by her bed either as she had grown accustomed to over the last twenty-eight years. A twinge of fear rose up her spine. She walked to the door to the outside chamber and opened it.

Her heart melted in terror as she saw the Praetorian guards outside of her sleeping quarters, waiting for her. The ten soldiers were in full ornamental battle gear and had no thoughts of allowing this girl to get away after the emperor's instructions. The emperor had a habit of decimating soldiers he surmised were not loyal. The practice consisted of killing one in ten men in a unit in order to ensure discipline. No, they would not let her get away. In fact they would enjoy this task.

"No, please, I can explain!" she shrieked and dropped to the floor, sobbing. Denouncing a virgin for incest against the state was a serious offense. The emperor used this opportunity shrewdly to blame this treason for his recent failures in battle and deflect blame from himself.

The guards picked her up off the ground and tied her hands behind her back. "It is too late for you; it has already been decided. You have broken your vow to Rome and the Temple of Vesta." Julia screamed in horror, as she knew what was awaiting her. She wailed as they dragged her by the hair from her bedroom and out of the College of Virgins. The pain added to her fear but was nothing compared to what she was about to experience.

The soldiers carried her to the cobblestone road outside the temple grounds in the center of Foro Romano. The crowds had already gathered, as the word had spread fast of what was to happen. This was even better entertainment than the gladiators in the Coliseum. No one wanted to miss the show. They were used to the emperor providing routine ghoulish spectacles to divert attention from their miserable and declining living conditions. The crowd was excited.

First she was whipped by a thick cane fifty times. The back of her white clothes became stained with blood. She was close to unconsciousness but hoisted onto a funeral cart and tied to a stake emanating from the center. A pail of cold water was thrown in her face to wake her up in order to enjoy the procession. Soon she was again screaming in horror and shock at what was happening. She had done nothing wrong! She had been true to Rome; she no longer even had any sexual feelings. Those had vanished long ago. She had been true to her oath. But today it didn't matter.

The cart wound its way through the center of the ancient city and slowly made its way outside of the massive walls to the place where the dead were buried, to the Campus Sceleratus, a small rise near the gate. Roman citizens lined the streets to witness the spectacle. Some were empathetic and sad, others were enjoying the cruel procession and used the occasion to start another decadent binge of drinking, drugs, and sex. The corruption of the Roman ethos was almost complete.

Julia had fainted with shock in the now hot sun. Her head hung limp on her shoulder as her body was supported by the cords strapping her to the pole. Soon the procession stopped. Another bucket of cold water from the nearby aquifer was thrown on her to wake her up. She looked up and hoped she had awoken from a nightmare but stood in shock as she realized it was not the case.

Her parents and family were screaming and crying behind a wall of soldiers protecting the executioner. The wealthy family, once close to the emperor's court, would now be banished. Their lives changed forever. They would be lucky to escape with their lives and would soon be making hurried plans to flee the city.

The soldiers untied her from the cart and dragged her by her bound hands to the hillside below. The ropes around her wrists and the rocks on the ground cut into her skin. She pleaded with them to listen to the truth, but they ignored her. They led her to an open tomb. They stopped in front of the crypt, and a Roman judge walked up to her and began to speak.

"You have broken your sacred vow to Rome. You will now accept the consequences of your pleasure of the flesh." At this point, Julia was too weak to protest. The soldiers walked into the crypt and placed an oil lamp on one of the slabs next to a decaying body. The air smelled of death, as one of the bodies was fresh from burial a week before. The soldier adjusted the wick and lit the lamp. Beside the lamp he placed a loaf of bread and a cup of water. Then they brought Julia into the tomb and pushed her to the floor. She was mumbling in an incomprehensible manner.

The rock was then moved to close the tomb. The last vestige of light twinkled out as the stone was rolled across the opening. Julia's voice was drowned out to the outside world, and her screams whispered like death on the wind.

No one noticed the richly dressed man standing in the crowd. If they did notice, they did not speak to him. The power emanating from his stature made it clear he was not to be spoken to, although no one recognized him as a local. He was dressed in a way which was foreign to Roman society, but it was obvious he was a man of sophistication. If they had noticed him, they would have seen he was smiling.

One hundred years later, Emperor Theodosius I extinguished the fire in the hearth of the Temple of Vesta as he proclaimed it inconsistent with Christianity, now the official religion of Rome. A few years later, Rome fell to the barbarians.

Chapter One

The words just wouldn't come. He sat at the wooden table on the old, rickety, spindle chair, his eyes attempting to focus on the laptop. The chair creaked beneath him. He had been meaning to buy something new to sit on, but in reality he just couldn't part with an old friend. It had been two months, and it was now like they were married. His body had molded a depression in the soft wood of the seat. How could he get a new chair? It was not going to happen.

The words still wouldn't come. Sometimes when he started to write, the words gushed out of him like a hot geyser, not tonight. He leaned forward on his elbows into the table and stared at the blue and white screen. He began to make out the pixels in the coloring. The table swayed slightly below him with the force of his weight and screamed in pain. Nothing. Today was not going to be the day he made progress. He just didn't know where to take the story. He needed inspiration.

So inspiration it would be. Rafe stood up. He wavered slightly as he stood and grabbed the table to steady himself. Glancing at the bottle of Chianti on the table, he could see it was half empty. Another dead soldier stood next to it. He should have been really drunk, but he wasn't. A couple bottles of wine a day saw to that. His tolerance was impressive.

Next to the first bottle was a picture of his kids. *I miss them, bad.* He turned away and pushed the thought out of his mind. *Actually, I'm the happiest I've been in a long time!* And he really was. The divorce had been over for six months now. Oh, of course he still had the raging phone calls, usually in the evening when she was drunk. But the good thing was he didn't have to be in the same room anymore. He had been taught not to hit a woman. So he had taken it—for years. It almost destroyed him. *No more.*

Rafe made his way confidently to the balcony door as he had been doing for two months now and opened it. The smell of the sea embraced him like a foggy morning. But it was evening. The sun was setting. The sun was setting on Venice. His balcony overlooked one of the canals meandering off the main drag. He could hear the taxi boats plodding along, their engines sounding like a growling lion. It seemed they never stopped. Venice, like any major city, had an extensive public transport system, except hers was on water. Luckily his building was tall, taller than most around him, and his view was spectacular from the top floor. He could see the clock tower in St. Mark's Square in the distance. The balcony was quite large, and he had started a nice collection of herbs and other plants growing in the Venetian sunlight and salty air. It was a heavenly location for a writer. It was just what Rafe needed to find some peace. He just wanted to write and be left alone for a while.

He turned and made his way back across the main room to the exit and started down the several flights of stairs, emerging along the canal into the evening shortly after. It was going to be a beautiful night. The gondolas for hire were taking full advantage of the perfect conditions for a moonlit cruise. After a few twists and turns and five minutes later, he arrived at his favorite restaurant along the water. The nightlife here was fantastic. It was when the real Venetians came out to play, avoiding the horde of tourists during the day.

Rafe sat at a table in the outdoor seating area, and soon a waiter was describing to him in Italian the specials for the evening. He understood only half of what was said but shook his head in agreement. "Surprise me," he said in English. The waiter smiled and started shouting instructions to the chef as only an Italian could do. Soon the food starting coming and didn't seem to stop.

In spite of all his problems, and besides missing his kids terribly, Rafe really *was* happy. This place enthralled him. *Maybe I will never leave Venice.* The authentic Italian seafood meal went down easy, and the courses were never ending. He found himself apologizing to the overweight, female proprietor why he just couldn't eat anymore. He paid the bill and spent an hour wandering through the passageways of the floating city, marveling at the history. The walk helped to digest his meal. The best thing to do in Venice was to get lost. He was on an island for God's sake, so it wasn't really being lost. He could find his way back eventually. But the pleasure was in finding a new street and meeting new people and watching the Venetians do their thing in their element, which was the evening. Soon he was doing just that as he found himself chatting with a local businessman who owned a tobacco shop. He enjoyed a fine cigar and became fast friends with the gentleman. *This is how I always find inspiration.*

Eventually, he decided to go back to his flat and get some writing done. Maybe the evening stroll would stir his imagination. He could still see St. Mark's in the distance, and he oriented himself to the bell tower. Soon he would come out near his home. Rafe found himself walking along a foreign canal in a neighborhood he did not know. It was strangely quiet and almost deserted. He enjoyed times like these, finding new places in his new favorite city, listening to the noises of the night. Rafe reveled in the fact he had no schedule and no one telling him what to do. He was completely in control of his own destiny, and he loved it.

He gazed at the ornately decorated palaces lining the canal and tried to imagine the history of the owners hundreds of years ago. The parties they threw, the beautiful women who lived there, all of this danced through his mind. He casually stopped along one such palace, long since deserted due to the mold creeping up to the upper floors from the constant flooding. He paused to take in the structure. It was times like these that inspiration came.

Venice was sinking. Slowly, very slowly, but sinking just the same. The buildings were constructed on wooden pilings sunk into the mud and clay centuries before. The foundations of the city's structures rested on this wooden support. The earth delayed the process of decay, but slowly these pylons were deteriorating. Artesian wells sunk in the early twentieth century to feed local industry were discovered to be adding to the structural problems, hastening the sinking of the city's support. As the city's elevation shriveled, the floods came more often and the damage grew exponentially. Many of the palaces along the waterways were deserted, or at least the first floor, due to mold and other hazards from the encroaching sea.

The population of Venice was now mostly older, as families with children had moved out long ago because of the safety hazards and expense of living on the island. The Italian government had spent hundreds of millions of euros to stop the decomposition of the city but could only slow not alter nature's course. Seawater had a nasty way of eating into a foundation over time that no amount of human intervention could stop. The future of Venice was in doubt in the long run. Today, however, Rafe enjoyed the scenery and wondered about the past.

As the evening light dimmed, out of the corner of his eye, Rafe noticed a strange glow emanating from the base of the palace. It was an orange, fiery color wafting through the water like cream in a coffee. He walked over and looked closer. The strange, colored light angrily turned bright red and then was gone. He shrugged and kept walking. *Must be the wine.* Darkness set in for the rest of his trip home.

His head hurt but not too much. His body was used to the alcohol. He was terribly thirsty however. The sun was peeking through the venetian blinds and stabbing him in the eyes. He awoke but didn't want to move. This was his favorite part of the day. He could just lie in bed until he couldn't lie there anymore. Rafe reached for a glass of water on the nearby table and downed it quickly. Then he closed his eyes. *I wonder what time it is. But, I don't really care.*

Rafe Savaryn was a world traveler. He loved exploring different civilizations, new and old. He wrote books about those experiences and taught history at a small Ivy League school in the northeastern United States. He especially loved European history. His family had emigrated from the Ukraine during the previous generation, and he still felt he had roots in Eastern Europe. "If you don't know history, you're doomed to repeat it," he always told his incoming classes.

Rafe spent several months a year in different, far-off corners of the globe. Previously he brought his family, but on this trip he was alone, due to the divorce. He enjoyed finding places that no one in the West knew much about, places that experienced a deep history that had been lost to the ages. Learning about the past gave Rafe great pleasure, as it helped him understand the present. This was the secret of his books and why he had become such a successful writer.

One of his earliest memories as a child was running across an expansive, open terrazzo with a large statue in front of him. He was obliviously happy and his mother was chasing him from behind, calling for him. He remembered her explaining to him about the statue and how important it was to history. Rafe ignored her and kept running. He remembered reaching the statue and seeing a tall, bronzed man riding a horse. As he grew older, he had always wondered where that place was. *Perhaps that is why I'm always searching and writing, hoping one day I'll run into that statue again. Perhaps that is where my curiosity for the past began.*

But his favorite place in the world was Italy. And his favorite city in Italy was Venice. To live among the houses where the Italians had fled from the barbarians during the Dark Ages after the fall of the Western Roman Empire, and built a city on the marshy islands, was heavenly for him. He felt as if he dined with da Vinci when eating among the locals in the late evening. He reveled in the atmosphere. Today was going to be no different.

An hour later, his side began to ache from lying in bed, and Rafe sat up, throwing off the matted sheets. He walked to the balcony and once again threw open the doors. He breathed in the sea. It was a daily ritual he enjoyed. He checked on his herb garden and then walked to the bathroom. Rafe took a short, cold shower to revive himself, quickly dressed, grabbed his laptop, and headed out of his flat and down the stairs. Today was the perfect day to write. He enjoyed the exercise as he strolled through the waking city. Soon he was sitting at a table in St. Mark's square, the tourists and the pigeons milling all around him. The ideas came and he began to write.

Hours later he came up for air. His eyes burned from staring at the screen, and his wrists ached from typing. He had finished five thousand words. *Quite the productive day if I do say so myself! Almost makes up for the horrible writing day yesterday.*

Rafe was in Venice to write a novel, a novel about the Renaissance. *And what better place to do that than here?* He looked at the bell tower rising forcefully high above all of the other structures. He remembered as a child his parents had a painting of St. Mark's Square hanging over the fireplace. He always wondered where the place was. Now he knew and he was sitting here. He experienced a form of déjà vu.

The sun was now slowly heading toward the horizon, and the light began to fade. Shadows made their way across the cobblestone. The tourists began to make their way to the boats. The dueling jazz and classical music orchestras across from each other in the plaza took turns playing to the locals and the tourists left on the island for the evening. The scene was magical. Rafe ordered his first drink of the night, and his mind wandered off into the past laid out before him.

He was jerked back to reality when he heard a young, female voice ask, "Is this chair taken?" Rafe, startled, turned to face the owner of the voice on his left side. A young woman not more than thirty sat next to him and signaled to the waiter for a drink. "I'll have what he is having," she said as the waiter arrived.

Rafe raised his eyebrow and said, "Very confident of you."

"You want me to stay, don't you?"

Rafe looked her over. She was young, thin, beautiful, and elegantly attired in a little black dress, her dark shiny hair rained down around her shoulders and framed her delicate face.

"I think I do," Rafe responded, a smile creasing his lips. She was of Italian descent he guessed, and her accent was deadly attractive. "Well this is a pleasant surprise," he added.

"I like to be spontaneous." They chatted about nothing for fifteen minutes or so, and Rafe smiled, as she was quite witty.

The first drinks went down easy, and Rafe signaled for another round. She began to speak but stopped as the drinks arrived and waited until the waiter was out of earshot.

"Do you like me?" she asked coquettishly.

"Yes, I do."

"Well, here's the deal. You take care of me and I'll take care of you."

"I figured it was something like that. The oldest profession?"

"I like to call it being a courtesan. I only work with high-end clients."

Rafe thought about it for a minute. The full moon shone down across the square, illuminating the pearls around her dark neck. *What the hell.* He reached into his wallet and pulled out several five hundred euro notes. Money was not a problem for Rafe. He was a very successful writer. He slid the notes across the table. "Will this do?"

She smiled. "I'm yours for the night."

They walked hand in hand through the darkened alleyways, occasionally stopping for a bite to eat or a drink at the many restaurants and bars dotting the landscape of Venice. Rafe enjoyed having some company for once. It had been several months since the divorce, but it had been years since he was happy with a woman. He realized he had been missing female companionship. *Rational, happy, fun companionship that is.*

Soon the hour was very late. He stopped in a darkened doorway that was indented several feet into the building, providing a very private space for exactly a moment like this. He pulled her close and felt her young, toned body under her dress. Her full breasts pressed against his chest. She kissed him and threw her arms around his neck, pressing her body into his. They embraced for several minutes.

They both were startled by a loud splash. Rafe looked up across the canal to the palace on the other side. Even in the dark, he could see the mold making its way up to the second floor like something out of a drive-in movie. The light from the full moon covered the water in a milky glow. Rafe looked around and realized he was at the same spot where he had seen the fiery water the night before. He walked over to where he had heard the object hit the water and looked around. He heard another noise below as the water was disturbed. He then realized part of the upstairs balcony of the abandoned palace was crumbling and pieces of stone were falling into the water.

They both peered into the water where the stone had entered as the ripples emanated from the entry point. Slowly an orange, fiery glow like the flame of a candle appeared as a small circle and then grew, spreading across the water like an oily flame. "What is that?" she gasped.

"I don't know, but I think I saw it last night as well." He leaned over the canal to try to get a better look. The mist turned a raging red and then vanished as quickly as it came.

"That's the weirdest thing I've ever seen!"

"Yeah, it's really weird. I just don't have any idea what it is!" Rafe stared a little longer and then turned away. "Come on, let's go."

They walked silently the short distance to his flat, climbed the stairs, and entered his studio. He kicked off his shoes and went out on the balcony.

"Do you have a name?" he asked.

She walked out with him and looked out over the moonlit city. "Cecilia. It's an ancient Roman name. I like it. My parents did good. And I know your name is Rafe." His eyes widened. "Don't worry, I always check out my clients."

She walked over to him and began unbuttoning his shirt. When she had finished, she pulled his shirt out of his trousers with his belt still buckled, exposing his stomach and chest. Her delicate hands caressed him.

She lightly kissed his chest then looked up at him. "You know next time, you're going to have to let me bring a friend." She bent lower to where her mouth was a couple inches above his belt buckle. "We both should be licking right here." Her tongue touched his skin. Rafe closed his eyes.

When Rafe awoke, the bed was a wreck but she was gone. *Oh well, at least it was worth the money. I feel like notching my bedpost or something.* He rose from the bed, looked out the balcony, and went through his morning ritual. This time, however, there were blackened clouds in the distance billowing down from the sky. He could hear the thunder and see the occasional flash of lightning. The storm was moving fast towards Venice, and he could see the people below scurrying to bring their things inside before the rain started. Just like that, the rain started pouring down in buckets. The clouds were violently churning and spewing thunder and lightning. He barely had time to shut the balcony doors.

Just then, the door jerked opened to his flat. Cecilia walked in carrying a tray with two cappuccinos and some pastries and fruit. She had changed into another knee-length sun dress from the small bag she had been carrying. "Breakfast is served!" she said. She looked out the balcony door windows. "Wow, that's an angry storm!"

"I thought you were gone for good."

"No, just thought I'd be nice to you and serve you something to eat. You were so gentle with me last night. And I had another thought. I thought maybe I'd just stay with you a while. You know, get to know each other. I like you."

"I'm not paying you any more money."

"I'm not asking, am I?"

'I guess not."

"You can just buy food. What do you think? I'm a kind of spur-of-the-moment person anyway."

"Well, I had fun last night that's for sure. So stay a while. But don't bug me, I've got to write."

"Yes, I know. The famous writer." She put the food down on the table next to his computer, walked over to him, pushed him back on the bed, and pulled her dress over her head. "You can start writing in thirty minutes."

She lay next to him with her head on his shoulder, her dark hair pushed up into his face. "You smell good but I need to write now, if I have any energy left."

"Of course, I'm not stopping you."

"So what's a nice girl like you doing what you're doing?"

"Aahhh, the big question. Well I'll tell you. I don't do it very often but I need money, and it's an easy way for me to get it. Capiche? I'm a perpetual student and I have to eat. Can you understand? Plus I like to travel, buy things, and meet interesting people. Does that make sense? I hope so, because it's the truth."

"Sure, it makes sense. I'm not judging you. I took you up on your offer, didn't I? Everybody has a price. What are you studying?"

"I'm an expert on the Roman Empire. Soon I will be rich and famous. I hope anyway. I give a lot of speeches now around Italy on the subject already. I'm quite the intellectual, believe it or not."

"I'm impressed! Maybe you can help me with parts of the book I'm writing."

"See, I knew you'd want me to hang around. Tell me about yourself, cowboy! I mean I know you are a famous writer and everything. I read an article that you'd be spending some time in Venice, writing your next novel."

"So that's how you found me? Ha. I remember that article. I was angry they wrote it. The guy caught me at a cocktail party in New York and presto, off-the-record comments show up in print. Well, what do you want to know?"

"Here's my big question. Why are you alone?"

"Let's save that question for another day."

Chapter Two

Rafe sat typing away at his laptop at a cafe overlooking the water near the fish market in Venice. It was a beautiful day and he was thrilled to be alive. He loved to sit in the middle of the crowds and imagine how life would have been centuries before. He tried to visualize how the remnants of the Roman population fled the barbarian advances from the north during the Dark and Middle Ages and took refuge in the marshlands off the coast, slowly building up the city over hundreds of years to become a major economic power, the most powerful city-state in all of Europe at one point.

The doge, or duke, as the Venetian leader was called, ruled the Adriatic as a major naval power during the early second millennium, building and operating thousands of ships and training accompanying crews. Venice even threated the Eastern Roman Empire at one point, sacking its capital Constantinople and occupying hundreds of Islands along the Adriatic coast, creating her own Latin empire. It wasn't until Christopher Columbus discovered the New World and opened up alternative trading routes that the power of Venice began to wane. A long and costly war with the Ottoman Empire served to irreversibly force her into decline. Venice was also an important republic during the Renaissance. She flourished as an independent city and patron of the arts until Napoleon Bonaparte conquered her in the late eighteenth century. The city-state became part of the Kingdom of Italy in the late nineteenth century.

The smell of fresh fish dominated the air, and sea gulls soared overhead in endless patterns, diving to pick at the discarded carcasses of the fish as they were cleaned and thrown in the trash bin near the alleyway. This was not a touristy area of the city, although some did come here to see the local atmosphere. The boats off-loading their catch came and went, and the market was bustling in the midday heat. Rows and rows of all different types of seafood were on display, nestled in a thick bed of ice. The Venetian women combed through the offerings, trying to find the best selection of fish to feed their families. Occasionally a tourist would wander into the market and request a picture with the mounds of whole specimens piled in the containers of ice, only to be shooed away by the market proprietors. This place was for selling fish, not for catering to tourists. Speed was of the essence to sell the entire catch, as the shelf life of a dead fish was limited before its freshness could not be guaranteed. Rafe tried to transfer all of the activity around him to his novel. There was nothing like seeing and describing events real time. It was a favorite technique of his.

There was not a cloud in the sky. It was as if all the rain the day before had washed away any hint of bad weather. He spent the previous day writing and was again quite productive. Cecilia had left him alone and let him work. That is until the sun began to set.

Rafe looked up from his computer as the memory of her body

returned. *Wow,* he thought. A sea gull swooped down over him and caught him by surprise. The scene was transposed immediately to his book. He attempted to capture the motions of the bird's wings, the sound of its call, and the aggressiveness of its attack while it attempted to acquire a scrap of seafood. He couldn't type fast enough.

It was late in the afternoon, and Cecilia had gone back to check out of her hotel and get her things. *I guess she's staying with me for a while. I'm probably crazy but what the heck.* He decided he had written enough for the day and paid his bill, downed the last bit of wine in his glass, stood up, and headed back towards his flat. He checked his word count and noticed he had only written a couple thousand words for the day. However, the prose was high quality and Rafe was satisfied.

The sun was beating down hard, and it was a scorching day. The tourists on the main drags were out in force, and he could hardly make his way down the crowded thoroughfare. It was no use. He decided to take an alternate route away from the well-traveled, popular parts of Venice. Soon he was on an alleyway along a canal devoid of people. Rafe felt relaxed and happy. *That's probably because of Cecilia. How strange is that? In other words, what the heck are you doing? Well, it feels right. One day at a time, I guess.*

A group of sea gulls again flew overhead making a god-awful racket, shattering his bubble of self-absorption. It was like they were trying to get his attention. He noticed they began to circle over the canal a few meters ahead. Instantly Rafe realized where he was. He was at the palace again. The palace filled with mold and the home of the fiery water. He walked to the edge of the concrete and peered into the green canal. There were no colors floating around, as it was the middle of the day and the light was bright. But there was something else. Rafe could see something attached to the wall of the palace about ten feet below the water, some type of symbol. He couldn't make it out, as the canal rippled on the surface and clouded his field of vision.

He felt drawn to this place. *I'm supposed to be here—how strange,* he thought to himself. The images below the surface beckoned him. He desperately tried to make them out through the trembling, murky water. It was no use. He sat on the edge of the canal and let his legs dangle over the side, frustrated, trying to decide what to do.

Screw this. Rafe looked around to his left and right and confirmed the alleyway was deserted. He quickly pulled off his clothes to his boxer shorts, looked around to check once again, and dove into the water. The salt filled his nose and eyes, and they burned. He made his way down to the symbol on the palace foundation. The light became less and less visible as he swam about ten feet under the surface. He arrived eye level with the image and stared as long as he could as his breath ran out. Rafe felt curiously alone and not alone at the same time. He stared at the image of the lion's head atop a naked man's body. A snake was wrapped around the torso with its head sitting atop the lion's mane. The body held two keys in its hands. The image had four wings protruding from his back, and a lightning bolt flashed across his chest. Rafe stared at the picture hewn into the stone. The lion's mouth was open and was terrifying. He needed to surface but he was held there. Something wouldn't let him go. He felt at home. Like the image was familiar, but he had never seen it before, of that he was sure. He felt light-headed and his lungs were bursting. *Surface you idiot!*

At the last moment, he used what strength he had, flailed his arms and legs, and kicked to the top. As his head broke the surface, the air exploded out of his lungs, and he sucked whatever molecules of oxygen he could manage. After several moments of regaining his breath, Rafe slowly swam overhand to the edge and pulled himself to the side of the canal and onto the walkway. He was exhausted. He lay there for several minutes, recovering. The image of the beast was still in his mind. Rafe could have sworn as he swam away that the face had turned red and blood oozed from the mouth of the lion.

It was some time before Rafe returned to his flat. The experience had jarred him. He had walked around the city for a couple hours, trying to make sense of it. But nothing made sense, he was confused. He opened the door to the flat as the sun began to set and the light was beginning to disappear from the balcony glass doors.

"I was beginning to think you wouldn't be coming back," said Cecilia as he noticed her lying on the bed. She was dressed in nothing but a terry cloth robe. Yet, his mind was still foggy from what had happened.

"I really don't know what just went down" Rafe responded.

"What do you mean?"

"As I was walking back from the fish market after writing there for several hours, I ended up back at the palace where we saw that strange glow in the water a couple days ago. It was the weirdest thing! I felt drawn to it, like I was supposed to be there. Then I saw something under the water mounted onto the stone foundation. It was an image of some sort. I couldn't see It clearly, so I dove in, just like that."

"Just like that?"

"Yeah, really strange, isn't it? his writing table and pulled a blank piece of paper from a notepad. He began to draw. A few moments later, he handed the drawing to Cecilia. "Like this."

She gazed at the rendering. "I've seen this before!"

"Where?"

"I don't know. I can't remember. But I know it has something to do with the Roman Empire. Can I use your phone?"

"Why?"

"I know someone who will know what this is." She took the phone from his hand as he held it out for her. She dialed.

"Fernando, hello, it's me Cece from Rome. Yes, it's good to hear from you as well. Listen, I don't have time to talk, but I want to send you an image, okay? I'll scan it and email it right over. I know I've seen it before but I can't place it. It's of ancient origin. Thanks, luv! Call me back at this number, okay?" She clicked off the phone and spoke to Rafe. "We've helped each other out from time to time. He is head of archeology at the Maritime Museum in Barcelona. I know he will be able to tell us what this is." She placed the image Rafe had drawn on the scanner and hit the button. The device began to hum. Shortly after, she logged in to her email from Rafe's computer and fired off the image. "Now we just wait!"

"Should I be jealous?" *Because strangely I am.*

"I'm not even going to answer that," she responded with a smile.

Embarrassed for asking, Rafe changed the subject. "So tell me more about your work."

"I have been studying archeology at Sapienza in Rome for some time now. There is always another class to take. I've focused on Roman society. It's quite fascinating. Several months ago, I was asked to give a speech on the subject at a diplomatic event in Rome. I have some contacts in the corpo diplomatic, and they were having an event, so they asked me to speak. I thoroughly enjoyed it. Since then, I've received multiple invitations. My talk is on why Rome collapsed. I'm even going to be paid for my next speech in Florence in a few weeks. It seems I've found my calling."

"Then I guess you won't have to hit up strange men in St. Mark's Square?" He regretted saying it right after the words left his mouth. He could see the look of shame sweep over her face.

"Hey a girl's got to survive, right? Anyway, I don't see you complaining. It takes two to tango."

"You're right. As I said, I'm not judging." There was an awkward silence between them. The phone rang unexpectedly and loudly, thankfully shattering the uncomfortable moment.

Cecilia answered. "Pronto, aaah Fernando, let me put you on speaker phone, okay? And speak English? I have a friend here as well." She placed the phone on the rickety table after hitting the speaker button.

"Where did you find this?" asked Fernando.

"Let's just say we found it in Venice," Cecilia responded.

"Well, I would like to know more at some point about where. It's a symbol of an ancient religion that we don't know much of anything about. It was very secret. It's called Mythraism. We do know it was mostly contained within the Roman legions. It seems they built underground temples wherever they garrisoned. The origins of the religion are unknown. There is some evidence that links it back to a god worshipped in Persia thousands of years ago, but the links are not definite or clear and mostly theory. There are a few symbols associated with this movement. This is one of them. The other primary one you see consistently is a statue of a soldier slaying a bull with a spear. Beyond that, we know nothing."

"But why would we find it here in Venice?"

"That's a good question, as Rome was long gone when Venice was built—that's why I want to know more. But today is your lucky day. I suggest you come to Barcelona. We've been doing some more excavation under the naval dockyards here in part of the museum space. We've found more of the old Roman city of Barcinoi. I suggest you come here and look immediately."

"But why do we need to do that?" she asked. "Can't you just tell us what you've found?"

"Because day after tomorrow, we are opening a two-thousand-year-old, underground temple we believe to be Mythraic."

After buying the plane tickets rather quickly, Rafe found himself leading Cecilia into the Venice evening to find a quaint, out-of-the-way restaurant for the evening meal. They made small talk for a while, enjoying a fine bottle of Pinot Noir and reveling in the discreet privacy of the small establishment. The place was tucked into the side of a fifteenth-century military structure; previously the area had been used as an armory or something of that sort. It was perched about ten meters above the canal below and the associated walkway beside it. It was one of Rafe's favorites. He had stumbled upon it rather accidently one day while wandering around the city. It was quite remarkable he had found his way back so easily, as it had been weeks ago. Cecilia was enchanted.

"You are full of surprises!" she gushed. "This place is marvelous."

"Yes, I thought you'd like it."

"So you never answered my question."

"What question was that?"

"Why are you alone?"

Rafe said nothing for a while then spoke. "It's a long story."

"We've got all night."

"I'm divorced. My ex-wife and I got married rather quickly. You see it was the great sex and all. Four months. Can you believe that? I was living high on Wall Street and she was my goddess. I was her hero. We turned out to be neither. Anyway, right after we were married, one evening, she became really violent. She totally changed. It was as if she wanted to get married quickly to not let me know what she was really like. It was scary. I mean like glazed-over eyes, talking in tongues kind of scary. Really dangerous stuff. I had no idea what to do. I was taught growing up you never hit a woman. But she was physically, emotionally, verbally, you name it, out of control abusive. It got worse and worse. And I took it. I took it for years. A year later, our first child was born. That was interesting, since the sex had stopped completely after we tied the knot, except for the occasional roll in the hay once a year. Anyway, things went downhill even more after that. I tried to keep her happy but it was no use. She was unhappy inside. Only later did I find out about the abuse she suffered as a child. Well, some years later we had another child, trying to put a happy face on this disaster, but it ended anyway. The divorce was final six months or so ago. I was literally exhausted. I needed a break. So here I am."

"Quite the story. I'm sorry for you."

"Don't be, as I told you, I'm the happiest I've ever been in my life. I wake up with a smile on my face every day. When you've been through what I've been through, to have a chance at a somewhat normal life is breathtaking. And, I have my kids. They are my sun, my light. I cherish them. My little girl is only four years old. She is a blessing."

"Do you miss them?"

"Of course, I talk to them almost every day, but this sabbatical has done wonders for me. I'm a new man. So, what about you? No boyfriend, sugar daddy?"

"Oh, there have been a few, but no one special right now. I dated a member of parliament for a couple years, went to all of the fancy parties and official events. But it got boring after a while. It was all about him, not about me. So I broke it off. Now I just focus on my career and traveling, when I can."

Rafe's phone rang. "Hello?" he answered.

"Hi, Daddee!"

"Hello, my little princess, how are you today? I miss you very much!"

"Daddy, I learned a fun song at school today. Do you want to hear it?"

"Yes, of course, my darling." Rafe listened as his daughter sang Twinkle Twinkle Little Star. His heart melted.

Dreams of the Negev

A Short Story

by

L Todd Wood

The siren started slowly but built quickly to a nerve-wracking crescendo, waking him rudely as the sunlight started to peak through the cheap, nylon curtains in the room. In an instant, he had a question to answer. *Do I go to the shelter or not?* The loud booms of the Iron Dome missile protection system shook the windows as the missiles found their targets, rockets fired at Tel Aviv by Hamas. *It will all be over soon anyway, no time,* he thought as he dug deeper into the covers for protection from the in-coming artillery. He was right. The sirens stopped momentarily. But, now he was awake. Groggily, he tried to piece together what was going on.

He knew he was in Israel, Tel Aviv to be exact, covering the war as a journalist. He had arrived several days earlier from the other conflict de jour, Ukraine, to report on the violence between Israel and Gaza for multiple news outlets. Having just turned forty years of age, his body no longer quickly adjusted to time differential when heading east. Although the sun was now brightly thrusting through the window shades like daggers as the Israeli day was starting, his body told him otherwise. *Sleep,* it said. He tried to get deeper into the covers. The cheap air conditioner in the low budget hotel had been running all night as the thermostat was broken. Even though outside the temperature hit thirty-five degrees Celsius, in his room, it was closer to fifteen.

He opened his eyes briefly. The sheet on the bed unfortunately had been kicked down to the foot of the mattress, he only had the synthetic bed cover over him and it wasn't enough. He was cold but he dared not move. The faux orange tile on the floor of the hotel room contrasted oddly with the tapestry hanging on the wall. It reminded him of a black velvet, hanging Elvis art piece one could buy on the side of the road in Memphis. The decor of the room was dated in the early seventies; however, everything was functional. He was thankful for that. Even the internet worked. The sounds were getting louder now outside as the stream of cars passing by his first floor window increased and the pedestrians walking towards the beach a few blocks away talked incessantly. Several dogs were barking. The same faded jeans and white shirt he had been wearing for a week sat rumpled on the floor. A have drunk glass of red wine stood next to the empty bottle on the bed stand. *Another dead soldier,* he thought.

He had been dreaming, that he was sure of. Lingering feelings of crime, pain, hurt, evil, abuse, and even death drifted through his clouded mind. He never knew why but those things were always there, subtly below the surface. But he also had a vague sense a door had opened somehow. It was as if he awoke too soon, that he was missing something, something important. That always happened when he wrote. He had stayed up late into the night, putting the finishing touches on a story. Writing was a gateway for him. It always led to other things. The dream was no exception.

For some reason, he also remembered it was his brother's birthday. His brother had passed away as a toddler when he was only four years old himself, from cancer. He had vague memories of him crawling across the floor to him on a yellow, ribbed, carpet. They weren't really memories, just flashes, snapshots of time stored somewhere in his brain. *Strange the things that enter your mind sometimes. I really am cold.* But, he didn't care and drifted off back to sleep.

He awoke to the sound of cats fighting. It was a horrible, screaming sound, like that of a woman being repeatedly stabbed. He rolled over in the sand and stared into the eyes of one of the felines who had walked over to him, searching for food. The cat's eyes were glowing a deep yellow, like that of the devil himself. He turned away, preferring not to return the animal's stare. His friend was dead, that he was sure of. There was no more breath coming from the body. He had taken a machine gun round to the chest. *At least I don't have to listen to the sucking chest wound anymore,* he thought to himself. The date was April 18, 1917 and he was in the Negev desert and he was cold. His body was shaking, attempting to ward off the chill. He relaxed as the sun began to peak over the horizon. *At least I will die warm,* he thought. The Gaza offensive had not gone well for the British. The Ottoman's were too dug in and well-defended. The Brits could not break their ranks. He had been cut off with his friend from the rest of his unit and they were pinned down behind a burnt out tank. As his friend was now dead, he was alone. When the light was strong enough for his enemy to see him, he would be finished off with artillery. *The only thing left to do is write. At least I can enjoy the remaining few hours of my life.* Writing was the only thing that brought him joy. He took out the few pieces of paper from his breast pocket along with a pencil and began scratching out a few sentences. For some reason, he wanted to finish the short story he had started a few days earlier. It was important for him to do that now. Even though it was highly likely no one would ever even know it existed, much less actually read it.

The birds were now circling overhead. They smelled the death of his friend. The blood had coagulated underneath him in the sand, creating a dark stain that had hardened in the heat. It was amazing how animals could pick up the scent of death from a long way away. The realization that they would pick his bones as well forced him to write faster. He was almost done. It was a love story, a story of two lovers reunited at the end of their lives. It would not be a story he would experience himself. He had long stopped thinking of his girlfriend. He had not seen or heard from her in over a year. His letters had not been returned. *I guess it's better that way than to get a Dear John letter,* he pondered.

He glanced up and saw the sun was now fully above the horizon and he was starting to sweat. *I don't have much time now!* He wrote faster. Eventually he peered over the small sand ridge in front of the hulk of the tank and saw activity on the other side. The artillery tubes were getting ready to fire. One of the spotters noticed his movement near the tank and he saw them point in his direction. *It won't be long now.* He wrote the final sentence, then turned on his back and stared up at the sky. *Strange, the birds have gone.* He heard the thump of the guns as the ground shook underneath him. *Goodbye,* he thought.

"Nyet!" he heard the man shout. His eyes fluttered open briefly and he realized he was still in the room under the nylon cover. The sun was now high in the sky outside as it was now midday in Tel Aviv. He could smell the Mediterranean ocean two blocks away in spite of the air conditioner, which was now straining to operate in the increased heat of the Israeli sun. A television was blasting in the lobby a few yards away with news of the war.

The Russian family had moved into the room next to him several days before and were very loud. They fought often, waking him from time to time as he slept. They were having another argument. He felt sorry for the daughter, who spent her time in the lounge chair in the hallway outside the hotel room, desperately trying to pass the time on her iPhone. She was approximately sixteen and had long blond hair and deeply tanned skin. She would be a real heartbreaker in a few years. There were many Russian Jews in Israel. The Jewish State was a natural place for Russian tourists to visit relatives who had immigrated after the fall of the Soviet Union. It had given him a chance to practice the language he had learned in college.

He thought about getting up, but his eyes still stung with fatigue. He rolled over, adjusted the pillows and drifted back to sleep.

The sentry stood on the ramparts of the fortress overlooking the desert sixty miles to the East of Gaza, and watched as the Roman siege ramp came closer and closer. Soon they would be here and he would die. There were only a few hours left. He had come to peace with his upcoming death. In fact, the whole settlement had decided to commit suicide before the Romans entered the fortress. The year was 74 A.D. and Caesar was angry. The Jews had revolted against Rome across the Levant and he wanted to make an example to the rest of the Roman Empire out of this last group of holdouts, perched upon the mountain in their citadel called Masada.

The world needs to remember us after we are gone, thought the sentry days before. For that reason, he had been writing every evening for the past two weeks, detailing the progress the Romans were making to overtake the fort and he's people's reaction to their upcoming death. He hid the scrolls in the temple, hoping they would be found long after the Romans had left.

He was one of the ten men chosen to kill all of the others. The job had been horrifyingly devastating to his soul. However, he had kept an eerie calm as he butchered men, women, and children alike, cutting their throats and letting the blood drain out of them. Now there were only the ten men remaining. Soon they would kill each other and the last man would kill himself. He was going to be that man. He would slit his own throat. But not yet. He had to finish the story first, the story of the Jewish revolt against the Romans. He wanted the world to know.

Below him, a thousand Jewish slaves, prisoners of war, helped build the ramp which the siege tower was now being slowly pushed upward. It had been three years that his people had held out in hope that somehow they would be spared. But alas, it was not to be. *They will be here in a few hours.* The siege tower pushed forward, inch by inch. He could see the eyes of the Roman soldiers, eager to ravage the population of the fort as they got closer.

He went to the temple with the last scroll and hurriedly wrote details of his last day. He wrote of his wife, whom he had killed. Her body lay in his home, where they had shared a wonderful last three years together. His children lay in bed with her. He wept for his youngest daughter, only two, when he had to cut her throat. As her father, he felt it was his duty to make sure she died fast. She had no forewarning of what he was about to do. He and his wife had made sure none of the children were aware of their fate before he took them outside one at a time and put the sharp knife under their chin. He wiped away the tears. There was no time for that now. He would see them again soon enough.

A bell rung. The hour had arrived. He picked up his sword and walked out into the sunlight to kill his friends.

The siren went off again. This time, he knew he was going nowhere. He had been in deep REM state and his body did not move. He didn't even open his eyes. Before he drifted back to sleep, one thought crossed his mind, *I wonder how you sleep in war? Or can you ever sleep?*

He dreaded seeing his parents, although the trip home was uneventful. Soon he was being led into the chamber where they waited for him. He straightened his spine, adjusted his clothing, and tried to look his best. Then he walked in.

"You are late," said his father. "We don't like to be kept waiting."

"You look terrible," said his mother. "Where did you get those clothes?" She motioned for a servant and barked, "Have him fitted immediately for some new clothing!"

"I tried to look my best," he said to deaf ears. The courtroom looked the same. There were lines of people from all over the galaxy waiting to see the Emperor and Empress. They did not have much time to speak to one of their children.

His mother softened a bit. "Come here and give me a hug," she cooed. "I have missed you. How is boarding school?"

"It's hard work," he replied, avoiding his mother's grasp. Two could play this game. "I'd like to see my brother."

"Your brother is out administering the outer planets. You will see him soon enough. When you have finished your studies! And only then! Since he left you he has been extremely busy, while you take your time finishing your degree at the university! What do you have to say for yourself?" his father boomed, the courtiers squeamishly trying to tend to his every need as his anger rose.

"I am working hard to finish father," he replied softly. "Well your writing has improved somewhat I have to say, judging from the reports the school has been sending me. Keep up the hard work and you will earn your rightful place on the throne. But it will not come easy." It was then he noticed a beautiful girl sitting to the right of his father. She had been very quiet but was looking at him strangely, with almost a morbid curiosity. His father saw him staring at her. "Ahhh," he said. "Meet Svetlana. When you return, she will be your queen. It has been decided. In ten years."

He looked at the girl and she turned her gaze to the floor and would not look him in the eye. But she had stolen a glance at him and the one thing he noticed was her deep blue eyes, as large as the ocean.

She was strikingly gorgeous. A terrible fear he had never felt washed over him. "I suggest you finish your studies and hurry home to her," his father added with a smile on his face.

"Yes, father," he softly replied and backed out of the grand hall of the palace as his parents were distracted with other matters of state. His mind was spinning with apprehension. The girl was a new twist to his life, one that he hadn't expected. Quickly, his thoughts returned to his parents. *At least that is over. They will forget about me for a while.* The empire had to be administered. It had been that way all of his life. On the voyage back to the university, he started again to write. *Maybe I'll make up a good story for Svetlana,* he said to himself.

The knocking came softly at first. He thought he was dreaming or was the knocking coming from the room next door? But soon he realized someone actually was knocking on the door to his room. "Yes," he answered softly? The chill in the air made him shiver and he worked up the nerve to reach for the sheet and add to his covering on the bed.

"Do you want the room cleaned?" a man's voice asked.

"No, not today, thank you," he responded as loud as he could muster as he groggily sat up in bed.

"Ok, thank you." The voice was gone.

He pulled the covers off and threw his legs over the side of the bed. He reached for one of his shoes on the floor and threw it against the switch on the wall that controlled the air conditioning. Blissfully, the cold air stopped spewing from the device mounted high on the wall. He decided to get up anyway. He slipped on his jeans and the white shirt, splashed some water on his face and opened the door to walk outside and get some coffee in the lobby of the hotel.

As he walked to the front area of the establishment, he heard the Russian man next door screaming some profanity. He didn't understand the word, but understood the meaning. It wasn't a nice comment to his wife. He shook his head in pity for the man's family. It was then he noticed the girl was still there, busily typing on her iPhone as if she could make all of the bad energy from her parents go away. As he walked by, she looked up at him and smiled. She had huge blue eyes.

MOTHERLAND

L Todd Wood

To the IDF girl in uniform I met on the bus from Tel Aviv to Jerusalem, and her grenade launcher.

"Our fatal troika dashes on in her headlong flight perhaps to destruction and in all Russia for long past men have stretched out imploring hands and called a halt to its furious reckless course."

Fyodor Dostoyevsky, The Brothers Karamazov

To me belongeth vengeance, and recompence; their foot shall slide in due time: for the day of their calamity is at hand, and the things that shall come upon them make haste.

Deuteronomy 32:35

Prologue

The Northern Lights have seen queer sights,
But the queerest they ever did see
Was that night on the marge of Lake Lebarge
I cremated Sam McGee.
(Robert Service, "The Cremation of Sam McGee")

Captain Richards was tired and cold when he arrived at the squadron. It was a dark, winter night, and the moon reflected off the frozen ground as he shut off the jeep's engine and doused the headlights. It was the type of cold you could only experience in the Yukon, the type of cold that when married to the desolate wilderness seemed not only to chill a man but frighten him with loneliness. He sat for a moment in the warm air, pondering what the temperature was outside. *I wish I was in my bed...* In a few seconds, with no heater running inside the vehicle, the cold air attacked feverishly through the hardtop. *Time to get out?*

He was issued the vehicle a few days before—one of the perks of becoming a flight commander. The army was giving him more responsibility. He took it and asked for more. He had been in the Alaskan territory half a year now, although it seemed like much longer than that. Europe was far away in his memory. He was not sure if that was a good or bad thing.

Finally making up his mind, he exited the vehicle, pulling the parka hood over his head. Still, the wind allowed the cold to seep into his bones as he walked the short distance to the entrance of the wooden, military World War II structure that housed the fighter squadron, the same type of building he had seen at many other army posts around the world. The white stuff on the ground crunched under his mukluks. At least the snow had stopped a few hours before. The door slammed as he closed it behind him, and he allowed himself a few seconds to savor the warmth inside. The chill radiated off his army-issued parka but slowly faded away.

Earlier in the night, he had been in a deep, dead man's sleep when the phone started ringing. Still drained from the previous day's brutal mission, he deserved the day off. But he could not say no to a full-bird colonel; even the famous Captain Richards, the war hero, could not do that. So he had dragged himself out of bed and put on a pot of coffee in the small kitchen provided in the Post Officers' Quarters.

He was not complaining, you understand. After a year fighting the Luftwaffe with the Royal Air Force in the United Kingdom, he appreciated the small comforts the army could give him. His own kitchen was definitely a perk, but he felt guilty about it. The faces of his long-dead friends at the hands of the Third Reich were still fresh in his mind. He hoped with time they would fade as well like the cold on his parka. The faces haunted him.

The squadron duty officer on the phone said it was urgent. Hell, things were always urgent in this war. *Hurry up and wait* was the soldier's mantra. Richards couldn't imagine what was so urgent that he had to be woken after flying for over fourteen hours back to Elmendorf Field in Alaska outside of Anchorage. But he was here, doing his duty. He always did his duty. The cold and the lack of sleep seemed inconsequential when compared to his comrades-in-arms still fighting on the Continent. No, Captain Richards wouldn't be complaining, not tonight. He would simply do what was asked of him. That's what soldiers do.

Warmed temporarily, Richards strolled purposefully through the hallway to the operations center, the boards in the floor creaking as he stepped on them. Maps of the local terrain adorned the walls along with pictures of wrecked aircraft. There were hundreds of wrecks in Alaska. Usually the hulks were left to rot in the tundra, which was dotted with these rusted remnants of flying machines, their stories long lost to the world. Richards could imagine the scene a hundred years from now, when some young pilot would fly over the carcass of a B-24 and wonder about the story behind it.

Soon he was facing the duty officer, a green second lieutenant who was lucky enough to get assigned here after being commissioned, rather than sent to the front in Europe or to some island, fighting the Japanese. He handed Richards a folded piece of paper. "Here are your orders," said the lieutenant. "The colonel says this mission is very sensitive and of the highest priority. This aircraft needs to get to Provideniya as soon as possible."

"That's it?" asked Richards. "Just another milk run to Provideniya? That's why the colonel got me up in the middle of the fucking night after a long mission yesterday? I guess he forgot about crew rest," Richards snarled sarcastically.

"No, I didn't forget," said a gruff voice behind him. "But this is war, and all the rules go out the window. Don't you know that, Captain?"

Richards turned. *Shit*, he thought to himself. "Good morning, sir! My apologies." Richards managed a crisp salute despite his fatigue. It was returned promptly.

"It's not a good morning, it's the middle of the fucking night," said the colonel. "And yes, it's urgent. This satchel has to be in Provideniya by close of business today. TODAY! You understand? Not tomorrow but fucking today!" He handed Richards a brown, leather, official-looking satchel. The briefcase was locked with a padlock containing a rotating dial key. "It's not me giving the orders on this one. This comes from way above my pay grade. It came from Washington. So yes, I gather *it's rather important*, but I have no idea why. Just get it there today so they get off my ass about this one, okay Captain?"

Richards took the satchel and slung it over his shoulder. He unfolded the flight orders and scanned them quickly. "Looks straightforward enough, sir. I'll get this done and be back in no time."

"Thank you, Captain. That is all. I know I can count on you. That's why I woke you up." Richards saluted again. The colonel again returned the salute, smiled as he turned, and left the operations room.

The lieutenant spoke again. "Weather is waiting for you at the tower. I got them up for you as well. Yes, it looks like this should be a milk run. Only some cloud cover to deal with, sir,"

"Thanks, LT. Have a good night, whatever is left of it. I'll report in once I'm on my way." Richards left the room and walked back towards the entrance to the squadron and the cold. The wind bit him in the face as he made his way back to the jeep. *At least the heater is warmed up,* he thought to himself as he turned on the engine. He drove the short distance to the flight line and the aircraft parked not far away.

Elmendorf Field sat on the edge of Cook Inlet, the natural port that fed the Alaskan territory. It was an effectively protected harbor, and the town of Anchorage sat behind the inlet, snuggled up against the Chugach Mountains rising above it. The mountains were covered in snow. However, in a few months when spring broke, the snow line would move higher up the slope, leaving only the peaks covered in white. In the summer, the reverse happened. As winter approached, the snow would slowly move down the mountainside day by day. The locals called this phenomenon *termination dust,* heralding the approaching winter.

The aircraft were lined up along the runway, over twenty-five of them; a few of them had their shark-painted snouts snarling at the cold. Most of them, however, were devoid of more than just the basic markings, as if they were meant for someone other than the Army Air Corp, and they were. The Russians would of course repaint them when they were delivered.

The maintenance chief met him at the tail number that was selected. "She's fueled and ready to go, Captain," he stated nonchalantly as he arrived. The man completed this task hundreds of times a week. It was nothing special.

"Thanks, Frank. I appreciate you getting up in the middle of the night as well. Seems to be something about this run that's very important to someone."

"Yeah, that's what I heard. They wanted this plane ready to go and quick. I did my part. Now it's your turn, Captain. Have a good flight. I guess you'll be taking the DC-3 back again from Nome. Fly safe."

Richards returned the sergeant's salute and climbed into the cockpit, situated himself, started the engine, and taxied out to the runway. The massive rotary engine growled loudly at the night. It was comforting to him somehow. It seemed to him he was the only thing moving in the dead of winter. After receiving clearance for takeoff from the lone man in the tower, he gunned the throttle and shot down the runway.

A half hour later, he was approaching the Alaskan mountain range and Rainy Pass to the West. The tundra flew by beneath him, and he allowed himself some enjoyment flying two hundred feet off the ground as the dawn broke behind him and illuminated the world ahead. The caribou darted and jinked below him, attempting to get away from the strange, terrifying noise. A smile of peace crept across his face. To tell the absolute truth, he had fallen in love with Alaska. The vast wilderness, the wildlife, the freedom, they all captivated him. He was home. He could feel it.

The mountains surrounded Anchorage like a bowl and protected the small town from the more severe weather Alaska had to offer in the interior of the territory. Rainy Pass was a natural opening that wound through the peaks and provided a way to navigate through the chain without crossing over the top of the twelve-thousand-foot mountain range. Over the millennia, the river had cut somewhat of a canyon through the mountains.

Richards had stowed the satchel in the small, rear cargo compartment of the Curtiss P-40 Warhawk. The United States had been providing fighter and other types of aircraft to the Soviet Union for over a year now to fight the Nazi war machine, which had invaded Russia and threatened Moscow, although now the Soviets had begun to push them back. It was important to keep the pressure on the Nazis from both east and west.

The Lend-lease program had been very effective, enabling America to delay her entry into the conflict. The effort primarily aided the United Kingdom, but the Soviet Union benefited as well. American manufacturing provided over twenty percent of Soviet aircraft used on the Eastern Front. Ships, trucks, tanks, and other heavy equipment were all provided in massive quantities. At the end of the war, the material was supposed to be destroyed under American supervision, but only a small percentage was actually disposed of in this manner. Much of the equipment the Soviets would use for decades after Hitler was no longer among the living.

Richard's squadron's purpose, in addition to providing an air defense capability for the Alaskan territory, was to ferry these aircraft to Russia via the Bering Strait. Provideniya was the small village and military installation on the eastern coast of Siberia, where he landed the planes to deliver to the Russian Air Force.

The lieutenant was right, the weather was not bad. The only issue was a cloud layer at eight thousand feet. He would have to navigate beneath the cloud cover through the pass, as he didn't have oxygen available on this flight, and it was filed as a VFR, or visual flight rules, hop.

He enjoyed flying under the clouds through the mountains. It was like driving a Formula One through the narrow streets of some European city. Rainy Pass was the only one he had not flown since his arrival in Alaska. He had been offered and had accepted an initial orientation flight through each of the other passes, a daylight trial run with an instructor in a two-seater. However, Richards was a very experienced pilot and he was not worried. *Just a milk run*, he thought to himself.

He entered the throughway with plenty of space above him before the cloud cover thickened and created a dark ceiling, which he had no desire to penetrate. The clouds were ominous. The weather was actually slowly becoming worse than the boys at the field had forecast. There was ice in those clouds, ice that could seriously degrade the aerodynamics of his aircraft. Ice was dangerous in Alaska. If you allowed enough of the stuff to accumulate on the wings of your plane, you stopped flying. That was not a good thing to say the least. It was as simple as that. If you could avoid it, you did. *I'm not going near it*, Richards thought.

He focused again on making his way through the pass. Initially, the canyon was wide, and the walls on either side seemed far away and harmless. However, as he approached the halfway point through the mountain chain, the pass started to narrow and the cloud ceiling began to drop, providing him less room for error.

"Not such a milk run after all," Richards mumbled to no one as he twisted and turned through the tangled canyon. He had slowed the fighter plane to just above stall speed to give him more maneuverability to negotiate the now very small passageway through the ominous mountains on either side of him. Sweat dripped down his forehead, stinging his eyes. The turns were becoming quicker and more dangerous; they required his utmost attention and skill.

I wish I'd had that orientation flight.

The visibility was dropping. The clouds above him were now emitting a thin veil of mist, which froze to his windshield. He continued to twist and turn and slowed the fighter even more, as much as he dared. The stall warning horn was buzzing in his ear. He wondered how much more time he had before he exited the mountain range on the other side and could relax on his way to Russia.

Where's that damn map? he thought to himself. His eyes quickly searched the cockpit, and he caught a glimpse of a sheet of paper on the floor under his left foot. The map had fallen down from where he had placed it on the left control panel. His aggressive flight maneuvers through the pass had seen to that. Richards took a chance and leaned down briefly to get the map, taking his eyes off the passage in front of him.

It was a mistake that cost him his life.

When he looked back up, his squadron commander's words came back to him with frightening clarity. "You've got to watch Rainy Pass; there is a twist halfway through that has killed many a skilled pilot. It's an S turn that is deadly if you're not careful."

The snow was falling hard now and the cloud cover was dropping fast.

As he glimpsed through the ice-covered windshield of the plane,

Richard's heart skipped a beat when he saw the mountain passageway turn ninety degrees to the right and completely block his path. The adrenaline coursed through his veins as he jammed the stick to the rear, cut the throttle, and slammed the control to the right to slow the aircraft in hopes of executing a wingover maneuver back down into the pass to the right. He forced the stick against his thigh so hard he broke several blood vessels in the process as the chemicals in his body created superhuman strength in the face of death. The Warhawk pitched up, and then the right wing dipped and she sank back down into the canyon. Out of the corner of his eye, he actually saw several carcasses of other aircraft lying in waste on the hillside that had made the same mistake. It was literally a graveyard of plane crashes.

He smiled as he thought he had made it and briefly imagined drinking vodka with the natives that evening at the local watering hole in Provideniya. However, as he banked harder to the right to fall faster back down into the pass, his right wingtip caught the mountainside and violently cartwheeled the aircraft down onto the boulder-strewn, granite slope. Mercifully, Richard's head impacted the left side of the cockpit with such force, he immediately lost consciousness.

The Warhawk made several flips before coming to rest in a small creek bed about two thousand feet above the floor of the pass meandering below.

In a few minutes, the snow covered any trace of tail number USSR-9328.

Chapter One

May 5, 1996
Rainy Pass
Alaskan Mountain Range
75 Miles West of Anchorage, Alaska

Captain Connor Murray tried to make out the herd of caribou bobbing and weaving a thousand feet beneath him through the chin bubble in the cockpit at his feet, the noise of the aircraft frightening them into desperate action. There were hundreds of the huge animals tromping over the lush, green tundra floor of the Alaskan wilderness. The calves tried to keep up with their mothers. The racks of the males stood out among the herd. The craggy, snow-covered peaks of the Alaskan mountain range rose on each side of him as he guided the HH-3E Jolly Green Giant through the pass on the way back to Elmendorf AFB, located on the outskirts of Alaska's largest city, Anchorage. The aircraft and crew were returning from ferrying supplies out to a forward operating, fighter alert base on the outskirts of the western coast, where the F-15s intercepted Russian bombers routinely trolling along the outskirts of the Alaskan Air Defense Zone, probing American air defenses.

"Lots of calves out here today, they must have just been born," said Airman Thomas, the PJ, or pararescueman, on board as he leaned over the open ramp at the rear of the aircraft. The only thing holding him inside the cabin was a webbed gunner's belt chained to the floor and fastened around his waist. "Too bad we don't have a sling and a water bucket today to have some fun. I'd love to try and drop a load of cold lake water on one of those big bucks!"

"I wish I had my rifle. One of those calves would feed my barbecue for a summer," added Master Sergeant Wolf, the flight engineer and the senior enlisted man on board. Instead, he picked up a high-powered camera and began snapping photos out of the starboard door in the cabin of the large helicopter.

"Now, now, Sergeant," Murray chided over the intercom from the cockpit, "the tree-huggers wouldn't like that so much, would they?"

"No but my wife would!" Wolf responded. "Heck, I might even get laid if I brought home that much meat to freeze for the winter."

Murray chuckled to himself.

"We can fit a calf in the back, can't we, Captain? I promise no one at the base will find out! I'll clean up the mess. And I'll invite you over for a steak this weekend!"

"No can do, Sergeant," replied the young captain in mock frustration. "Against squadron policy and every rule in the book, you know that!"

"But we can call it a relief stop, Captain! And I do really have to go! You can put her down somewhere on the meadow I see out the starboard side."

"Can't do, Wolf. I'd lose my wings. I just made aircraft commander for God's sake. I'm sure the relief tube in the back will do the trick just fine."

"I can't seem to find it, Captain! All I can find is your helmet bag!"

"Ha. And I was just looking for someone to sit rescue alert this weekend!"

"Okay, okay, you got me! I guess I'll just have to suffer with no game meat next winter. Geez, I've been through this pass a thousand times and I've never seen so much caribou. Bagging one would be like taking candy from a baby!"

Murray's thoughts returned to flying as he skillfully piloted the giant helicopter through the twists and turns of Rainy Pass. Ahead, the famous S turn in the canyon loomed. This was Murray's favorite part, and he slowed the aircraft about seventy knots to give himself more maneuver room through the tight passage.

"The ice is melting early this year," stated Airman Thomas. "You can see much more of the canyon walls. Man, there's a ton of Dall sheep out this year too. Look at them all over the rocky slopes."

"What the hell...what is that? Tally-ho!" Wolf suddenly shouted excitedly into the microphone in his helmet. "Target three o'clock low!"

"Whatcha got, Wolf?" Murray responded.

"There's a wreck down there I've never spotted before, and as I said, I've been through this pass a thousand times. Can you go back, Captain? Seventy-five percent was sticking out of the ice. One of the wings was broken off."

"I'll try, Wolf, but this better be good!" Murray pulled back hard on the stick, reduced the power to minimum, and the giant helicopter shot nose up into the sky. There was no cloud cover so Murray had plenty of room. Once the airspeed had bled off, Murray kicked the right pedal and pulled the cyclic towards the window, and the nose of the Jolly Green Giant pulled to the right and back down into the canyon, executing a perfect yet violent one-hundred-eighty-degree wingover turn.

"Jesus, Captain, remind me never to give you a hard right turn unless I mean it!" said Wolf as he held on to the cabin wall for dear life.

"There she is!" shouted Thomas. "Holy shit, that's a P-40 Warhawk. Man is this our lucky day!"

"And she's just decided to show herself after all of these years," added Wolf.

"You're sure she's a new wreck?" asked Murray from the cockpit.

"Yes, sir. I am very sure. I know this pass like the back of my hand. I've never seen her before," Wolf responded.

"Okay, then I'm going to report it to RCC. Heading up! Jeff, why don't you take her while I get the HF dialed in. You have the controls."

"Sure, Connor. I have the controls," stated the copilot.

"I'm marking her on the map," added Murray.

Jeff Raines, the copilot, took hold of the collective and pulled back on the cyclic and added power. The powerful machine bolted upwards into the thin air, as she was light on fuel. Ten minutes later, the Jolly Green Giant cleared the top of the mountain range at twelve thousand feet. In peacetime there was a ten-thousand-foot limit an aircrew could not pass without an emergency. A new wreck sighting, especially one from WWII, caused Murray to waive that regulation as the aircraft commander.

"We'll just stay up here long enough to make the call, since we don't have oxygen," stated Murray to the crew. The craggy summits of the mountains spread out below him. Murray marveled at the beauty. A wonderland of mountains, ice and snow carpeted the earth as far as the eye could see. He dialed in the correct frequency to the HF, or high-frequency, long-range radio, clicked the switch on the stick, and spoke into the microphone protruding from his helmet and touching his mouth. "RCC, this is Jolly 26. Do you read, over?" Nothing. He tried again. "RCC, Jolly 26." The Jolly Green call sign was a holdover from the Vietnam War, where the large, green aircraft made history retrieving pilots shot down over North Vietnam.

A few seconds later, there was a warbled response. The HF radio used only high frequencies as opposed to UHF (ultra high) or VHF (very high), and the quality of the transmission was usually poor; however, the range was much longer.

The RCC, or Rescue Coordination Center, was a unit located at Elmendorf Air Force Base in Anchorage, which coordinated rescue assets throughout the Alaskan theater. Located in the basement of one of the command buildings, the unit was responsible for also cataloguing all wrecks in the Alaskan territory so as to efficiently use government resources if a crash was reported. There was no need to send out a rescue crew to a site the RCC knew was a previously reported wreck. There were thousands of wrecks all over the UHAE, or Unique Harsh Arctic Environment, "yoo-hay" as the locals called it.

Major Jordan was leaning back in his chair, sleeping, when the HF call came in from Jolly 26. He was overweight, lazy, and really bored with his job. When things were slow, he typically took catnaps in his chair behind the radio console. The assistants on the row in front of him couldn't see him. At least that's what he told himself. But he knew they made fun of him. *Probably the reason I've been passed over for lieutenant colonel,* he thought to himself. *I just have two more years to make it to twenty and a cush retirement.* It took him a moment to focus on the situation. Duty in the RCC tended to be long periods of boredom sometimes broken by quick periods of excitement, like if a crash was reported and a rescue launched. Jordan's dreams of his young, female airman assistant were rudely doused as the radio blared.

"Jolly 26, this is RCC, go ahead, over," she responded.

"RCC, we have uncovered a new wreck. It seems to be of World War II vintage and appears to be a P-40 Warhawk carcass. It's slowly melting free from the ice. Coordinates are as follows." Murray reported the latitude and longitude of the crash site as the airman feverishly copied. Major Jordan sat up in his chair, now alert.

"Ask them to investigate," he told her.

"Jolly 26, request you investigate, over."

"Roger, RCC, WILCO. Jolly 26 out."

Captain Murray took in the situation and made a decision. The HH-3E was originally a naval helicopter; in fact, the fuselage was shaped like a ship's hull for floating on the water if needed. The U.S. Air Force had adapted this machine for combat rescue missions in Vietnam a couple decades before, adding a ramp and air refueling capability. The problem was the engines on the Jolly Green Giant were not situated for high-altitude operations. They simply were not designed for hovering in the heat or where the air was thin. The good thing was that they were low on fuel, since they were almost at home base, having already drained the tip tanks on the long trek back from the Alaskan coast.

"Wolf, compute some power data for me. I want to see if we can hover over the site. The area looks relatively flat, so we should be in ground effect. I think the altitude is around two thousand feet. We are light, so I bet we're okay to go take a look."

"On it, sir!" Wolf responded. The flight engineer pulled out a metal-encased notebook and quickly made some calculations on a preprinted table and came to a positive conclusion. "You're right, Captain. We can hover. We'll even have some excess power. The bad thing is we'll only have about five minutes on station before bingo fuel back to base."

"Thanks, Wolf. Okay, crew, we're going down to take a look. You know the drill. If anyone sees anything dangerous, call a go-around. The escape path will be down and to the right into the canyon if we get into trouble, like losing an engine or something."

Hovering a helicopter was a delicate maneuver. Hovering an underpowered, twenty-thousand-pound helicopter at two thousand feet on a hot day was downright dangerous, a procedure not to be taken lightly. Murray set up the approach into the wind and left himself plenty of room to slide down into the canyon on the right if he had a power or control problem. Slowly, the huge machine reduced speed and lost altitude in a controlled manner. Three minutes later, they sat in a stable, fifty-foot hover to the right of the crashed aircraft. Murray concentrated on controlling the Jolly Green as the cabin crew scanned the crash site.

Wolf spoke first. "The left wing is visible as well as part of the rear fuselage. It's definitely a P-40, WWII vintage. Strange, none of the usual U.S. markings however. Just painted a dull green color. She's mostly intact. However, most of the right wing is still under the ice a ways back; must have broken off in the crash. Captain, can you slide back twenty? Maybe I can get a look at the tail number."

"Sure, Wolf, coming back slowly. Let me know when to stop." Murray eased back on the cyclic, added a little more power, and the giant helicopter moved backwards ever so carefully.

"Three, two, one, hold her there, Captain!" Murray did as instructed, all the while maintaining his reference to the ground through the chin bubble at his feet. "I got the tail number! It's USSR-6328. Shit, it's a damn lend-lease aircraft!"

Lieutenant Raines, the co-pilot, jotted the number down on his kneepad for future reference.

"We're bingo fuel, Captain, time to go!" said Wolf as he peeked into the cockpit between the pilots.

"Roger that, pulling power." Murray added power and dove off the side of the mountain to gather airspeed and to move from a hover into translational lift. Soon the power requirement was reduced as the airspeed over the rotor system increased, and he turned the aircraft back down towards the opening of the pass on the way to Anchorage. He could see the city's center buildings in the distance on the other side of the bowl, the Chugach Mountains rising behind them. Twenty minutes later, the HH-3E emerged from the western mountain pass and now was within range for normal VHF communication with Elmendorf.

"RCC, this is Jolly on Victor, can you read?"

"Roger, Jolly, what do you have for us?"

"We've got the tail number. I'll spell. Uniform, sierra, sierra, romeo, dash, niner, tree, two, eight." Murray spoke using the military phonetic pronunciation to prevent miscommunication over the airwaves, hence the strange pronunciation of three.

Major Jordan took the piece of paper with the tail number written on it from his female airman assistant, all the while noticing her slim waist and endowed chest. He smiled to himself at his coup of getting her assigned with him at the RCC. Jordan turned in his swivel chair to the computer console behind him. He brought up an internal search page and typed in the tail number and coordinates, expecting to see the crash site logged years before. The airmen jotted down the event into the daily logbook.

However, Major Jordan's attention was immediately fixated on the computer screen as the answer came back. He sat up straight as a board in his chair, and a low whistle escaped from his mouth.

IMMEDIATE OPSEC PRIORITY – NATIONAL SECURITY HIGHLY CLASSIFIED SITUATION. DO NOT COMMUNICATE ABOUT THIS INCIDENT OVER UNCLASSIFIED CHANNELS. IMMEDIATELY NOTIFY THEATER-LEVEL COMMAND UNITS. PREPARE TO SECURE CRASH SITE AS SOON AS POSSIBLE WITH TOP SECRET, SPECIAL COMPARTMENTAL CLEARANCE ONLY.

"Holy Mother of God!" Jordan muttered as he picked up the secure line to the wing commander.